Lecture Notes in Computer Science 13582

More information about this series at https://link.springer.com/bookseries/558

Sokratis Katsikas · Steven Furnell (Eds.)

Trust, Privacy and Security in Digital Business

19th International Conference, TrustBus 2022
Vienna, Austria, August 24, 2022
Proceedings

Editors
Sokratis Katsikas ⓘ
Norwegian University of Science
and Technology
Gjøvik, Norway

Steven Furnell ⓘ
University of Nottingham
Nottingham, UK

ISSN 0302-9743   ISSN 1611-3349 (electronic)
Lecture Notes in Computer Science
ISBN 978-3-031-17925-9   ISBN 978-3-031-17926-6 (eBook)
https://doi.org/10.1007/978-3-031-17926-6

This Springer imprint is published by the registered company Springer Nature Switzerland AG
The registered company address is: Gewerbestrasse 11, 6330 Cham, Switzerland

Preface

Recent computing paradigms, such as cloud computing, big data, and the Internet of Things (IoT), open new horizons to businesses, improving effectiveness and ensuring that businesses remain competitive in the global marketplace. These developments have occurred in a remarkably short time span, with technologies sometimes advancing too fast for society and governments to keep pace and leading to concerns around trust and the extent to which information security and user privacy can be ensured.

In answer these concerns, the 19th International Conference on Trust, Privacy and Security in Digital Business (TrustBus 2022), held in Vienna, Austria, on August 24, 2022, provided an international forum for researchers and practitioners to exchange information regarding advancements in the state of the art and practice of trust, privacy, and security in digital business. As in previous years, it brought together researchers from different disciplines, developers, and users all interested in the critical success factors of digital business systems. The conference program comprised two technical paper sessions covering topics in the areas of privacy, encryption, and security in IoT. The papers were selected by the Program Committee via a rigorous reviewing process (each paper was assigned to four referees for review) and six out of 15 papers were finally selected for presentation as full papers at the conference.

The success of this conference was a result of the effort of many people. We would like to express our appreciation to the Program Committee members, to the external reviewers for their hard work, and to the members of the Organizing Committee. Last but not least, our thanks go to all the authors, who submitted their papers, and to all the participants. We hope that you will find the proceedings stimulating and beneficial for your future research.

August 2022

Sokratis Katsikas
Steven Furnell

Organization

Program Committee Chairs

Steven Furnell	University of Nottingham, UK
Sokratis Katsikas	Norwegian University of Science and Technology, Norway

Program Committee

Cheng-Kang Chu	Institute for Infocomm Research, Singapore
Nathan Clarke	University of Plymouth, UK
Frédéric Cuppens	Polytechnique Montréal, Canada
Sabrina De Capitani di Vimercati	Universita degli Studi di Milano, Italy
Vasiliki Diamantopoulou	University of the Aegean, Greece
Josep Domingo-Ferrer	Universitat Rovira i Virgili, Spain
Prokopios Drogkaris	ENISA, European Commission, Greece
Jan Eloff	University of Pretoria, South Africa
Eduardo B. Fernandez	Florida Atlantic University, USA
Jose-Luis Ferrer-Gomila	University of the Balearic Islands, Spain
Simone Fischer-Hübner	Karlstad University, Sweden
Sara Foresti	Universita degli Studi di Milano, Italy
Jürgen Fuß	University of Applied Sciences Upper Austria, Austria
Dimitris Geneiatakis	Joint Research Centre, European Commission, Belgium
Dimitris Gritzalis	Athens University of Economics and Business, Greece
Stefanos Gritzalis	University of Piraeus, Greece
Marit Hansen	Unabhängiges Landeszentrum für Datenschutz Schleswig-Holstein, Germany
Ying He	University of Nottingham, UK
Pallavi Kaliyar	Norwegian University of Science and Technology, Norway
Christos Kalloniatis	University of the Aegean, Greece
Georgios Kambourakis	University of the Aegean, Greece
Farzaneh Karegar	Karlstad University, Sweden
Maria Karyda	University of the Aegean, Greece
Vasilios Katos	Bournemouth University, UK
Anne Kayem	University of Potsdam, Germany

Dogan Kesdogan	Universität Regensburg, Germany
Spyros Kokolakis	University of the Aegean, Greece
Stephan Krenn	AIT Austrian Institute of Technology, Austria
Costas Lambrinoudakis	University of Piraeus, Greece
Antonio Lioy	Politecnico di Torino, Italy
Javier Lopez	University of Malaga, Spain
Stephen Marsh	Ontario Tech University, Canada
Fabio Martinelli	IIT-CNR, Italy
Leonardo Martucci	Karlstad University, Sweden
Vashek Matyas	Masaryk University, Czech Republic
David Megias	Universitat Oberta de Catalunya, Spain
Chris Mitchell	Royal Holloway, University of London, UK
Martin Olivier	University of Pretoria, South Africa
Rolf Oppliger	eSECURITY Technologies, Switzerland
Pankaj Pandey	Norwegian University of Science and Technology, Norway
Maria Papadaki	University of Derby, UK
Andreas Pashalidis	BSI, Germany
Ahmed Patel	National University of Malaysia, Malaysia
Günther Pernul	Universität Regensburg, Germany
Nikolaos Pitropakis	Edinburgh Napier University, UK
Joachim Posegga	University of Passau, Germany
Gerald Quirchmayr	University of Vienna, Austria
Ruben Rios	University of Malaga, Spain
Panagiotis Rizomiliotis	Harokopio University of Athens, Greece
Carsten Rudolph	Monash University, Australia
Pierangela Samarati	Universita degli Studi di Milano, Italy
Ingrid Schaumüller-Bichl	University of Applied Sciences Upper Austria, Austria
Miguel Soriano	Universitat Politècnica de Catalunya, Spain
Stephanie Teufel	University of Fribourg, Switzerland
A Min Tjoa	TU Wien, Austria
Aggeliki Tsohou	Ionian University, Greece
Edgar Weippl	University of Vienna, Austria
Christos Xenakis	University of Piraeus, Greece

External Reviewers

Vaios Boulgouras	University of Piraeus, Greece
Cristòfol Daudén-Esmel	Universitat Rovira i Virgili, Spain
George Iakovakis	Athens University of Economics and Business, Greece
Sergio Martinez	Universitat Rovira i Virgili, Spain

Contents

PriPoCoG: Guiding Policy Authors to Define GDPR-Compliant Privacy Policies

Jens Leicht[1]([⊠]), Maritta Heisel[1]ⓘ, and Armin Gerl[2]ⓘ

[1] paluno - The Ruhr Institute for Software Technology,
University of Duisburg-Essen, Duisburg, Germany
{jens.leicht,maritta.heisel}@uni-due.de
[2] University of Passau, Innstrasse 41, 94032 Passau, Germany
Armin.Gerl@uni-passau.de

Abstract. The General Data Protection Regulation (GDPR) makes the creation of compliant privacy policies a complex process. Our goal is to support policy authors during the creation of privacy policies, by providing them feedback on the privacy policy they are creating. We present the Privacy Policy Compliance Guidance (PriPoCoG) framework supporting policy authors as well as data protection authorities in checking the compliance of privacy policies. To this end we formalize the Layered Privacy Language (LPL) and parts of the GDPR using Prolog. Our formalization, 'Prolog-LPL' (P-LPL), points out inconsistencies in a privacy policy and problematic parts of a policy regarding GDPR-compliance. To evaluate P-LPL we translate the Amazon.de privacy policy into P-LPL and perform a compliance analysis on this policy.

Keywords: Privacy policy · Policy language · General Data Protection Regulation · Formalization · Prolog

1 Introduction

Defining privacy policies is a difficult task, especially since the General Data Protection Regulation (GDPR) of the European Union [4] came into force. Using GDPR-compliant privacy policies is of much importance, because a violation can incur high fines. We present the Privacy Policy Compliance Guidance (PriPoCoG) framework that supports policy authors in creating GDPR-compliant privacy policies. Additionally, our framework can support data protection authorities in the evaluation of privacy policies.

PriPoCoG uses Prolog-LPL (P-LPL), our formalization of the Layered Privacy Language (LPL) defined by Gerl [5]. We use Prolog, a logic programming language, to formalize LPL and parts of the GDPR. This makes it possible to evaluate a privacy policy against a set of given rules and facts, for example, that Germany is part of the EU. Thus, our framework enables policy authors (and

S. Katsikas and S. Furnell (Eds.): TrustBus 2022, LNCS 13582, pp. 1–16, 2022.
https://doi.org/10.1007/978-3-031-17926-6_1

data protection authorities) to automatically check the GDPR-compliance of privacy policies. The full-fledged PriPoCoG framework will also support end users in understanding privacy policies and giving informed consent to data collection and processing. In this paper, we focus on supporting policy authors.

We provide some background in Sect. 2, followed by related work in Sect. 3. Next, we present an overview of our framework in Sect. 4. We discuss P-LPL in Sect. 5. Our evaluation of P-LPL using the Amazon.de privacy policy is presented in Sect. 6. Finally, we discuss our results and future work in Sect. 7.

2   Background

We give a brief introduction to the GDPR, LPL, and Prolog.

2.1   General Data Protection Regulation

The General Data Protection Regulation of the European Union (GDPR) [4] applies to any business that wants to interact with customers inside the European Union. It defines rules for a privacy-protecting interaction between end-users (data subjects) and service providers (data controllers). Data controllers may face substantial fines if they do not comply with the GDPR.

Data transfers to different data recipients inside and outside of the European Union are an important part of the GDPR, and we take these data transfers as an example for our formalization of GDPR requirements in Sect. 5. The European Commission may come to the decision that a third country provides sufficient data protection, implementing rules similar to the GDPR. Such an adequacy decision can be used by data controllers to demonstrate that their data transfers to third countries comply with the GDPR.

2.2   Layered Privacy Language

The Layered Privacy Language (LPL) framework by Gerl [5] is intended to cover the complete privacy policy life cycle to aid the protection of personal data by design.

LPL is a tuple-based language, and Gerl provides syntax and partial semantics for LPL, but computer-interpretable syntax and semantics are not available. The tuple-based definition of the language is well suited for our formalization using Prolog. For our formalization of privacy policies and GDPR requirements, we use an updated version of our extension of LPL [9]. The update to the LPL-extension includes updated data categories and purposes, according to the latest version of the Data Privacy Vocabulary (DPV) [10].

2.3   Prolog

We formalize LPL and parts of the GDPR using SWI Prolog[1]. Prolog is a declarative logic programming language that was standardized in the ISO 13211–1

---
[1] https://www.swi-prolog.org/.

standard in 1995 [7]. SWI Prolog is one of many implementations of Prolog, which adheres to the ISO standard.

Prolog programs consist of clauses, called rules and facts, where rules are expressed using Horn clauses. Rules consist of a *head* and a *body*, see Eq. (1). If the *body* of a rule evaluates to true the *head* is true, too.

$$head : - \; body. \tag{1}$$

The most important logical operators of Prolog are conjunction ',' and disjunction ';', which are used to express the bodies of the rules. The body of a rule consists of predicates, also called goals, which are calls to further facts or rules. A fact is a simple statement that states that something is true.

$$country(``US"). \tag{2}$$

Equation (2) is a fact that states that "*US*" is a known country. Prolog further distinguishes variables and atoms. Variables are named with a capital first letter, e.g., 'CO' in Listing 1 in Sect. 5. Atoms, that is fixed values, begin with a lower-case letter or are written as strings, e.g., 'intervention' (cf. Listing 2 in Sect. 5) or '"*US*"' in Eq. (2).

3 Related Work

Caramujo et al. [3] created a domain specific privacy policy language. This language allows the specification of computer-processable policies, similar to P-LPL. However, this language only performs consistency checks and is not capable of performing GDPR-compliance checks. Furthermore, instead of a new domain specific language, we use an existing declarative programming language to express the requirements of the GDPR.

Torre et al. [12] use artificial intelligence (AI) to check policies against the GDPR. However, their concept is limited to checking the 'completeness' of privacy policies. This completeness check does not check whether the regulations on permissible data processing are complied with. Furthermore, the usage of AI may not be a sufficient proof of compliance, as it uses a black-box approach that is hard to review. Lawyers or other experts can review P-LPL to assure that the checks actually represent the requirements of the GDPR.

Whilst our framework considers the creation of GDPR-compliant privacy policies as its main goal, Kiyomoto et al. [8] focus on the management of privacy policies and users' consent to specific processing of their data. Kiyomoto's privacy policy manager could be adapted for the use with P-LPL policies, to manage users' consent and privacy preferences.

Slavin et al. [11] describe a framework for the detection of privacy policy violations, which could benefit from computer-processable privacy policies, like the P-LPL policies created with our framework. P-LPL policies would render the usage of natural language processing (NLP) unnecessary, thus improving accuracy.

Yang et al. [13] use NLP to extract purposes from textual privacy policies. This concept could be adapted for our framework, to semi-automatically translate existing textual privacy policies into P-LPL.

PriPoCoG performs compliance checks as described in Sects. 4.3 and 5. These checks work on the information provided by the policy author and give hints when information specifically required by the GDPR is missing. Bhatia et al. [2] also perform incompleteness checks in their work. These incompleteness checks are not based on specific articles of the GDPR; instead, they focus on answering all relevant questions that a data subject or data protection authority may have. Our framework already checks policies for missing information that is required by the GDPR. However, extending these checks by requiring information that answers the questions identified by Bhatia et al. [2] may improve it.

4   Framework

We now explain the concept of our framework, and the procedure of creating GDPR-compliant privacy policies using it. Furthermore, we take a look at what articles of the GDPR our framework covers. Finally, we give a brief preview of the privacy policy editor that is part of our framework.

4.1   Overview

We support policy authors during the creation of privacy policies for their services. Our Privacy Policy Compliance Guidance (PriPoCoG) framework performs GDPR-compliance checks during the definition of a privacy policy, informing authors about potential compliance issues in the policy they are creating. Furthermore, data protection authorities can use our framework to perform compliance checks on service providers' privacy policies.

In Fig. 1, we present an overview of PriPoCoG. *Policy Authors* interact with a *Privacy Policy Editor* via the *I-PPE* interface. They can define privacy policies for their services, as well as check policies for GDPR-compliance. The policy editor retrieves compliance feedback from Prolog - LPL (P-LPL), our formalization of the Layered Privacy Language (LPL) [5]. We also formalized parts of the GDPR in P-LPL, which enables us to perform compliance checks on privacy policies. P-LPL is discussed in more detail in Sect. 5. *Data Protection Authorities* can use a different user interface, the *Privacy Policy Compliance Interface*, to check compliance of privacy policies.

Our framework can also be used to define policies that are used between data controllers and data processors to replace/enhance Service Level Agreements (SLAs). Furthermore, we plan to also support end users in the future. Thus, the PriPoCoG framework will be enhanced with further components and interfaces. Overall, we strive for a comprehensive support in handling privacy policies.

Since all compliance checks make use of P-LPL, privacy policies that shall be checked for compliance need to be expressed in LPL/P-LPL. However, since we provide a privacy policy editor and support policy authors in the definition

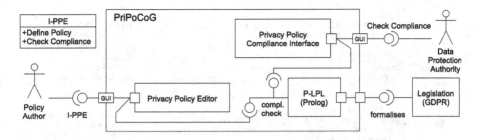

Fig. 1. Structural overview of the PriPoCoG framework

of privacy policies in P-LPL, this issue only exists when the aim is to check the compliance of existing textual privacy policies. This issue could be resolved using natural language processing to semi-automatically prepare P-LPL policies from textual privacy policies, as mentioned in Sect. 3. Generating natural-language policies from P-LPL, on the other hand, would be a relatively simple task.

4.2 Procedure

The definition of a GDPR-compliant privacy policy using our guidance framework consists of three steps, which we visualise in Fig. 2. All steps in grey boxes require manual work by the policy author. Step two is performed automatically in the background. Step three can be ignored when the compliance of the privacy policy was attested in step two.

1. Step (Manual). In the first step, the author defines a new privacy policy for a service, using our privacy policy editor, which we briefly describe in Sect. 4.4. Alternatively, the policy author can take an existing privacy policy and translate it into P-LPL, using the policy editor.

Input: The author needs to have knowledge about the system behaviour and business processes to be able to create a privacy policy for the system. If an existing policy shall be translated, this policy is required as input for this step.

Output: The policy created by the author is stored as a P-LPL policy.

2. Step (Automatic). In the second step, P-LPL performs a number of compliance checks, which we describe in more detail in Sects. 4.3 and 5. The feedback generated from these compliance checks is visualised in the privacy policy editor, to inform the author of any issues identified. Depending on the feedback, the result of the procedure can either be a GDPR-compliant privacy policy, if no issues were identified, or step three needs to be performed to achieve GDPR-compliance.

Input: A P-LPL privacy policy, which was either created in step one or modified in step three.

Output: P-LPL outputs feedback concerning the GDPR-compliance of the policy, which is visualised by the policy editor. The editor provides hints on how the author can improve the policy to achieve GDPR-compliance.

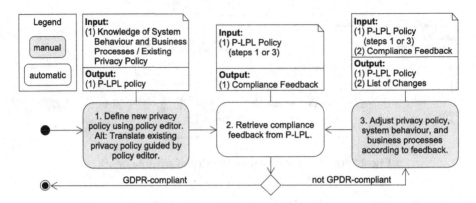

Fig. 2. Procedure of checking privacy policies for GDPR-compliance

3. Step (Manual). The policy author can now manually adjust the privacy policy in places where GDPR-compliance is not yet achieved. The feedback generated in step two shall be used to identify critical parts of the privacy policy that need to be changed.

Input: The privacy policy created in step one or previously edited in step three, and the feedback retrieved from P-LPL in step two.

Output: The policy adjusted by the author is stored as a P-LPL policy, which is again used as input for step two. Additionally, a list of changes is created to allow authors to track their changes.

4.3   GDPR Coverage

The articles of the GDPR [4] contain a wide range of rules that cover various aspects of data protection and are concerned with different stakeholders. Some parts are concerned with the behaviour of data controllers and processors, which must be expressed in a privacy policy. We formalized all those parts of the GDPR that are concerned with privacy policies and that can be observed by analysing the privacy policy of a service. The following list shows the 29 articles that are checked by our formalization:

$$5 - 7, 9, 10, 12 - 22, 27, 28, 31, 37, 42, 44 - 47, 49, 77, 88, 89$$

Article 5 – which our formalization covers to a large extent – describes the following principles of data protection. Principles that we cover with P-LPL are marked with (+), and principles that we do not cover are marked with (-).

Principles of the GDPR:

1. Lawfulness +, Fairness -, Transparency +   5. Storage Limitation +
2. Purpose Limitation +   6. Integrity, Confidentiality -
3. Data Minimisation +   7. Accountability -
4. Accuracy -

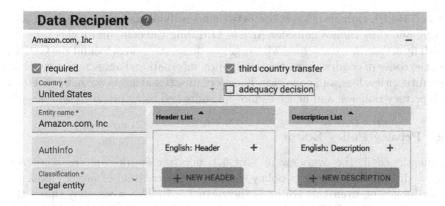

Fig. 3. Exemplary data recipient element in our privacy policy editor

Principles 4, 6, and 7 require knowledge about business processes and data management of the service (6 and 7), and principle 4 requires specific behaviour of the service provider, to allow data subjects to keep their information accurate. Such knowledge is not part of privacy policies and hence cannot be checked by our system. Principle 1 can be checked partially. By improving the GDPR-compliance of privacy policies, we support service providers in performing lawful processing of data. Principles 2, 3, and 5 are fully covered by P-LPL.

By using a formal privacy policy language (P-LPL), different ways of representing the privacy policy can be used, which improves the transparency of privacy policies. P-LPL policies can for example be represented using the privacy policy representation pattern by Gol Mohammadi et al. [6], to give a high-level overview first, whilst allowing data subjects to retrieve more detailed information if desired. Interested data subjects could then dig deeper into the policy by looking at a more detailed representation of the policy, or even a purely textual representation.

Articles 28 and 47 are special cases, where the privacy policy guidance does not directly cover these articles. However, P-LPL can also be used to express Service Level Agreements (SLAs) between data controllers and data processors, thereby supporting Article 28. The same holds for binding corporate rules, as described in Article 47, which can also be expressed in P-LPL. Articles 28 and 47 can, hence, be addressed using our privacy policy guidance framework.

The GDPR, overall, contains 99 articles, of which we cover 29. In the following we discuss some of the reasons for why 70 articles cannot be checked by P-LPL. Articles 1 to 4 contain introductory information and definitions, which do not need to be checked in the context of a privacy policy.

Article 8 is concerned with the data protection regarding information concerning children. Since age restrictions are not part of privacy policies, such checks are out of range of our framework.

If the processing does not require identification of the data subject, as described in Article 11, the data subject rights need not be granted to the data subject. P-LPL cannot differentiate between identifying and non-identifying processing and thus cannot consider Article 11 during the compliance checks.

Many articles are concerned with organizational structures behind the GDPR, e.g., the codes of conduct, certification bodies, international agreements, supervisory authorities, fees, and penalties. These organisational issues are not concerned with privacy policies, and hence cannot be addressed using privacy policies.

4.4 Privacy Policy Editor

As mentioned in the overview in Sect. 4.1, our framework contains a privacy policy editor, which supports policy authors in the creation of privacy policies. A small excerpt from this editor is shown in Fig. 3. The screenshot shows the form used to enter information regarding a data recipient. The information shown here is taken from our case study, which we present in Sect. 6. A number of other forms exist with a similar "look and feel".

5 Formalization: P-LPL

P-LPL is a Prolog program consisting of two parts.[2] The first part is a direct implementation of LPL. We take the basic version of LPL from Gerl [5] and, additionally, we consider our extension of LPL [9]. The extension further improves LPL's compatibility with the GDPR by, for example, adding a new classification for data recipients (i.e.,' Public Authority'), which allows us to identify data recipients that act as a public authority of a country.

Our implementation of LPL in Prolog only considers LPL's syntax and type checking. For the second part of P-LPL, we formalize those parts of the GDPR that are concerned with privacy policies and the information expressed in these policies, as discussed in Sect. 4.3. By way of Prolog rules, we define compliant and non-compliant combinations of values of a privacy policy.

This separation of concerns allows for an easy adaptation of P-LPL to other legislations. Compliant and non-compliant value combinations related to the GDPR can be removed or directly adapted to other legislations, and additional rules can be added if required.

The resulting Prolog code of P-LPL consists of 140 clauses and 371 facts, which cover a total of 1150 lines of code. The facts include lists of known country codes, as well as countries falling under an adequacy decision, which can also be updated easily, in case new adequacy decisions are formed by the European Commission. An example of a simple fact is the country code shown in Eq. (2) in Sect. 2.

In the following, we present some of the checks that are provided by P-LPL.

Third Country Checks are an example of compliance checks. They use the information concerning the data recipient to check whether the use of the data by this data recipient is compliant with the GDPR. The focus of these checks are the Articles 15 (2) and 46 (1) as well as Article 45 of the GDPR.

---

[2] Available at: https://github.com/jensLeicht/PriPoCoG.

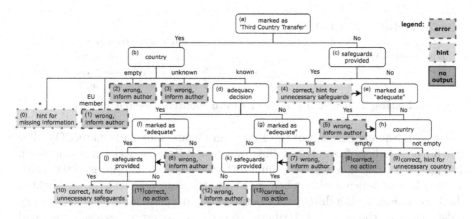

Fig. 4. Decision tree for third country transfer checks

Articles 15 (2) and 46 (1) require the provision of suitable safeguards if the third country does not fall under any adequacy decision of the European Commission. Article 46 (1) requires policy authors to state whether a third country falls under the European Commission's adequacy decisions. Article 45 further describes conditions that apply when an adequacy decision is available.

Third country transfer checks are implemented in *thirdCountryDR* clauses. The decision tree in Fig. 4 provides an overview of all 14 results (labelled (0) to (13)) of the ten *thirdCountryDR*-checks (labelled (a) to (k)). The decision points in the tree are based on the values of the data recipient as entered by the policy author in the form shown in Fig. 3. Relevant values are the third country transfer checkbox, the country the data will be transferred to, the safeguards (omitted from Fig. 3), the adequacy decision checkbox, and the internal information about existing adequacy decisions, represented as facts in P-LPL.

The decision tree includes three correct value combinations that are GDPR-compliant and consistent (green/solid border). Furthermore, there are four combinations that provide a hint for unnecessary or missing information (yellow/-dotted border), and seven combinations that provide detailed error descriptions for non-compliant or inconsistent content of the policy (red/dashed border). The hint for missing information (0) means that the policy lacks information necessary to check compliance of third country transfers.

Each of the letter-marked boxes (a) to (k) represents a decision point, which either leads directly to feedback to the author, or to further decision points. The decisions lead to feedback or to the conclusion that the data recipient is defined in compliance with the GDPR. The arrows starting from the **hint** and **error** boxes (4) to (7) visualize the fact that some messages can occur simultaneously. For example, an adequacy mismatch may occur simultaneously with missing safeguards, leading to multiple error messages being produced for a single data recipient.

Listing 1 shows the check that leads to result (6) in the decision tree, formalized in Prolog. The check determines whether a given third country data recipient falls

```
thirdCountryDR(N,TCT,CO,AD,_,P) :−
    TCT,                              % thirdCountryTransfer
    adequate(CO),                     % to an adequate country
    \+ AD,                            % not marked as adequate
    ...                               % inform author (error)
```

Listing 1. Prolog: Example of a third country consistency check

under an adequacy decision of the European Commission (adequate(CO)). For this purpose, a list of countries falling under an adequacy decision is consulted. The fact that a decision exists is compared to the attribute *adequacy* (*AD*), which is given by the policy author when specifying the data recipient (cf. highlighted checkbox in Fig. 3). If a discrepancy between the attribute and the given country is detected, i.e., the provided attribute is set to false, an error message is provided to the author.

Data Subject Rights. P-LPL uses a hierarchical structure to organize purposes for collecting and processing of data inside the policy. This structure makes use of a set of predefined purposes, which P-LPL uses to check some requirements of the GDPR regarding data subject rights (DSRs). Article 20 (3) of the GDPR, for example, states that the right of data portability does not need to be granted if processing is performed in the public interest or is performed by public authorities.

To model this case, we make use of the legal basis 'publicTask' to express that processing is performed in public interest. The legal basis is part of each LPL purpose, and it describes the legal bases for processing, according to Article 6 (1)(b) to (f) of the GDPR [4]. Official authorities are identified using the data recipient classification 'Public Authority'.

In addition to the restricted right of data portability mentioned above, Article 17 of the GDPR states that the right of erasure does not need to be granted for archiving purposes in the public interest. This exception is also identified by the legal basis 'publicTask', in combination with the given purpose 'ResearchAndDevelopment'.

Listing 2 shows another check regarding DSRs. If automated decision-making (ADM) is used, the DSR of human intervention must be present in the privacy policy. The check retrieves information about ADM from the given list of purposes (*P*) and checks if ADM is declared in the policy. If ADM is declared but the DSR *intervention* is not present in the policy, feedback to the author is provided.

Further Checks. In addition to the checks discussed in the previous sections, we also formalized the following checks:

Legal Obligations. When a processing purpose is defined with the legal basis 'legal obligation', stating that the data controller is legally obliged to collect and process some data, this purpose must be marked as *required*. Additionally, such purposes must be sub-purposes of the given purpose 'LegalCompliance', either directly or indirectly in the purpose hierarchy.

```
dsrCheck(D,_,P) :-                          % D = Data Subject Rights, P = Purposes
   admCheck(P,R),                           % retrieve ADM information from purposes
   length(R,L),                             % get length of retrieved ADM
   L > 0,                                   % at least one instance of ADM was found
   \+ memberchk((intervention,_,_),D),      % DSR 'intervention'
   ... .                                    % not present => output error message
```

Listing 2. Prolog: Automated Decision-Making; Data Subject Right 'Intervention'

Data Protection Officer. A dedicated data protection officer is only required in some circumstances, for example, when the data controller is a public authority. Using the classification 'Public Authority' in the context of the data controller, we can assure that a data protection officer is declared in the policy. If no data protection officer is declared, we provide feedback to inform the policy author about this compliance issue.

Policy Feedback. Depending on the error or discrepancy that was detected in the privacy policy, different types of feedback are provided to the policy author. We differentiate between **HINT**s and **ERROR**s. This clarifies which part of the policy must be changed to be compliant and where changes may improve the comprehensibility of the policy by removing unnecessary information or providing additional detail. An example of specific feedback is presented in Listing 5 for the case study in Sect. 6.2.

6   Case Study: Amazon.de

In this case study, we translate a real-life privacy policy into P-LPL. Since Amazon.com's privacy policy is not GDPR-compliant and is missing a lot of basic information required by the GDPR, we instead use the English version of the German Amazon.de privacy policy for our case study, which is intended to be GDPR-compliant.

6.1   Methodology

We extracted the latest version of Amazon.de's privacy policy [1] (last updated 2020-12-04) from Amazon.de's website. The case study focusses on the main privacy policy document and any linked content that provides information about the parties involved in the privacy policy (e.g., the data controllers). We do not consider external links to the cookie policy and further information regarding advertising partners since these contain a lot of outdated information. These outdated references to advertising partners would hinder us in the retrieval of necessary information about the data recipients. Furthermore, the privacy policy does not explain which tracking system may receive/handle what information.

After extracting the policy from the website, we started the translation process by first analysing the policy, identifying sections that do not contain any relevant information for our formal privacy policy. The main policy document

"Amazon Europe shares customers' personal information only as described below and with Amazon.com, Inc. and the subsidiaries that Amazon.com, Inc. controls that are either subject to this Privacy Notice or follow practices at least as protective as those described in this Privacy Notice."[1]

Listing 3. Excerpt from the Amazon.de privacy policy formalized in Listing 4

contains 4410 words, 1786 of which are not relevant for the specification of a P-LPL policy. These irrelevant parts of the policy contain redundant descriptions of already presented information or introductory texts.

Listing 3 shows a snippet from the Amazon.de privacy policy. We translated the sentence to the P-LPL clause shown in Listing 4. The P-LPL clause corresponds to the data recipient shown in the privacy policy editor in Fig. 3.

The first two lines of the formalization in Listing 4 contain most of the details of the data recipient: i) the name *"Amazon.com, Inc"*, ii) the data recipient is not a person, but a *"Legal Entity"*, iii) *""* and *"DataRecipient"* are internal values of P-LPL, iv) the Boolean values *true* and *true* state that this data recipient must be accepted to use the service and that a third country transfer is performed, v) *"US"*, *false* states that the country where the data will be processed is the United States of America, and that the country does not fall under any adequacy decision, vi) *[...]*, *[...]* marks natural language descriptions that are left out due to space constraints, vii) *SG* is a reference to the safeguards SG that are presented in lines three to seven. The safeguards contain a name *"Privacy Notice"* and natural language descriptions of the applied safeguards, for example, *"This privacy notice [...]"* , which can be provided in diverse languages.

6.2 Results

We manually extracted 100 data elements that are processed, which include six duplicate elements. In P-LPL, we were able to replace these duplicate elements with references, resulting in 94 distinct data elements.

Our formalization provides feedback concerning three out of a total of four data recipients, as Listing 5 shows. The privacy policy contains three data recipients with insufficient information regarding the country that the data will be processed in. This is caused by the generic way in which Amazon captured their

Listing 4. P-LPL: formalization of Listing 3

```
dataRecipient(dr1, (" Amazon.com, Inc", "Legal Entity ",   "",   " DataRecipient",
             true, true, "US", false ,   [...],   [...],  SG)) :-
SG=[("Privacy Notice",                                    % safe guards
   [" de ","  Datenschutzerklaerung "),(" en "," Privacy Notice ")],     % headlines
   [(" de ","  Diese Datenschutzerklaerung oder eine strengere ."),   % descriptions
   (" en "," This privacy notice or more protective rules ."]
 )].
```

HINT: third country data recipients should be provided with a specific country to be
   able to perform a compliance check: Third Party Sellers(PurchaseDelivery)
HINT: third country data recipients should be provided with a specific country to be
   able to perform a compliance check: Third Party Sellers(AmazonServices)
HINT: third country data recipients should be provided with a specific country to be
   able to perform a compliance check: New Business Owner(BusinessTransfers)
true.

Listing 5. Feedback for Amazon.de's privacy policy

data recipients: Amazon.com Inc, subsidiaries of Amazon.com Inc, third-party
sellers, and the new business owner (in case of a business transfer):

For subsidiaries of Amazon.com, it is unclear which entity may process the
data and whether these entities are based in countries outside the European
Union. Similarly, third-party sellers can be based anywhere on earth and end-
users have to reconsider data protection for any third-party seller they interact
with. P-LPL requires information regarding the destination country to be able
perform all compliance checks. Without this information third country transfer
checks are incomplete. However, it may make sense to leave out information
about the destination country: Since a business transfer is not planned in advance
and no information about the new business owner is known at the time of the
creation of the privacy policy, the country information should be left empty. In
case no information regarding the country can be supplied, the policy author
manually has to assure that this data recipient is GDPR-compliant.

The keyword **true** at the end of Listing 5 shows that the rest of our formalization was checked without issues. To ensure that the feedback is not caused
by mistakes during the translation of the policy into P-LPL, we double-checked
the formalization of all elements that were contested by P-LPL.

6.3   Performance

To gain insight into P-LPL's performance, we calculate average execution times
for 100 policy assessments. It is necessary to run the P-LPL formalization of
the same policy multiple times because of the non-deterministic nature of the
Prolog execution engine and the influence of other programs being executed in
parallel on the same computer. Non-deterministic here refers to the fact that
internal results may be generated in an arbitrary non-deterministic order, which
can influence the execution time. Non-deterministic here does not mean that the
results of the policy evaluation may vary from execution to execution, but instead
internal and temporary results may be processed in any order. The performance
evaluation was performed on a notebook with a quad-core CPU at 1.5 GHz and
16 GB system memory.

In order to gain insight into the influence of the size of textual parts of the
policy on the execution time, we carry out the evaluation on two versions of
the Amazon.de privacy policy. The first version contains headlines and textual
descriptions for every element in the policy. These are filled with 150 words
of placeholder text (Lorem ipsum . . .), to assure that each and every natural

Table 1. Performance evaluation (memory usage, policy size, and execution time)

Policy	Memory used	Policy size	Average time
Inflated	5.9 MB	456 KB	0.041 s
Original	5.2 MB	26 KB	0.034 s

language text field has a considerable text size. We name this policy 'inflated' in Table 1 above. The second version (original policy) contains many empty text elements that we were unable to fill due to missing information in the original textual policy.

As Table 1 shows, the average time to receive feedback on the Amazon.de privacy policy is 34 milliseconds. The difference between the inflated and the original Amazon.de policy is negligible with only seven milliseconds of increased processing time (20% increase), although the policy size is increased by 16 times the original size.

This evaluation shows that the Prolog implementation performs well on the case study. Even artificially inflated policies can be processed within tens of milliseconds.

Although the compliance analysis of a privacy policy using P-LPL only takes milliseconds, the translation of textual privacy policies, written in natural language, into P-LPL takes a significant amount of time. The translation of the Amazon.de privacy policies took several hours, since it first needed to be restructured, and only then the content could be expressed using P-LPL. The time consumed by the translation can be reduced by using P-LPL directly from the beginning of the policy definition process.

7   Discussion and Future Work

We presented our Privacy Policy Compliance Guidance (PriPoCoG) framework that supports policy authors in the creation of GDPR-compliant privacy policies. As discussed in Sect. 4.3, a substantial part of the GDPR can be covered by PriPoCoG. Limitations stem from the fact that we are working exclusively with privacy policies and do not have insights into the business processes of the data controller. We presented examples of our formalization of GDPR requirements in P-LPL, which is the backbone of our framework. Our evaluation using the privacy policy of Amazon.de shows that a compliance check for a complete policy only takes milliseconds. Hence, policy authors are not negatively impacted by the compliance checks.

In the future, we plan to improve the policy editor, by providing explanations concerning required information and descriptive compliance feedback. We are currently working on an extension of our framework to also support data subjects in the understanding of privacy policies. The extension will also cover the interaction between data controller and data processor, by using P-LPL policies as Service Level Agreements (SLAs). Privacy policy management is also an important problem that needs to be addressed in the future. Our goal is to provide a comprehensive privacy policy framework, supporting different stakeholders.

Acknowledgement. We thank Thomas Santen for his useful comments.

References

1. Amazon Europe Core: Amazon.de privacy policy (2020). https://www.amazon.de/gp/help/customer/display.html?nodeId=201909010&language=en_GB, Accessed 11 Jan 2022
2. Bhatia, J., Evans, M.C., Breaux, T.D.: Identifying incompleteness in privacy policy goals using semantic frames. Requirements Eng. **24**(3), 291–313 (2019). https://doi.org/10.1007/s00766-019-00315-y
3. Caramujo, J., Rodrigues da Silva, A., Monfared, S., Ribeiro, A., Calado, P., Breaux, T.: RSL-IL4Privacy: a domain-specific language for the rigorous specification of privacy policies. Requirements Eng. **24**(1), 1–26 (2018). https://doi.org/10.1007/s00766-018-0305-2
4. European Parliament, Council of the European Union: Regulation 2016/679 of the European Parliament and of the Council of 27 April 2016 on the protection of natural persons with regard to the processing of personal data and on the free movement of such data, and repealing Directive 95/46/EC (General Data Protection Regulation). Official Journal of the European Union L119, 1–88 (2016). http://eur-lex.europa.eu/legal-content/EN/TXT/?uri=OJ:L:2016:119:TOC
5. Gerl, A.: Modelling of a Privacy Language and Efficient Policy-based De-identification. Ph.D. thesis, Universität Passau (2020). https://nbn-resolving.org/urn:nbn:de:bvb:739-opus4-7674
6. Gol Mohammadi, N., Pampus, J., Heisel, M.: Pattern-based incorporation of privacy preferences into privacy policies: negotiating the conflicting needs of service providers and end-users. In: Proceedings of the 24th European Conference on Pattern Languages of Programs, pp. 1–12 (2019)
7. ISO 13221–1:1995: Information technology - Programming languages - Prolog - Part 1: General core. Standard, International Organization for Standardization, Geneva, CH (1995)
8. Kiyomoto, S., Nakamura, T., Takasaki, H., Watanabe, R., Miyake, Y.: PPM: privacy policy manager for personalized services. In: Cuzzocrea, A., Kittl, C., Simos, D.E., Weippl, E., Xu, L. (eds.) CD-ARES 2013. LNCS, vol. 8128, pp. 377–392. Springer, Heidelberg (2013). https://doi.org/10.1007/978-3-642-40588-4_26
9. Leicht, J., Gerl, A., Heisel, M.: Technical report on the extension of the layered privacy language. University Duisburg-Essen (2021). https://doi.org/10.17185/duepublico/74569
10. Pandit, H.J.: Data privacy vocabulary (DPV). Draft, Data Privacy Vocabularies and Controls Community Group (2021). https://dpvcg.github.io/dpv/
11. Slavin, R., et al.: Toward a framework for detecting privacy policy violations in android application code. In: Proceedings of the 38th International Conference on Software Engineering, pp. 25–36 (2016)

12. Torre, D., et al.: An AI-assisted approach for checking the completeness of privacy policies against GDPR. In: 2020 IEEE 28th International Requirements Engineering Conference (RE), pp. 136–146. IEEE (2020)

13. Yang, L., Chen, X., Luo, Y., Lan, X., Chen, L.: PurExt: automated extraction of the purpose-aware rule from the natural language privacy policy in IoT. Secur. Commun. Netw. **2021**, 1–11 (2021)

Closing the Gap Between Privacy Policies and Privacy Preferences with Privacy Interfaces

Stefan Becher$^{(\boxtimes)}$ ⓘ, Felix Bölz ⓘ, and Armin Gerl ⓘ

University of Passau, Innstraße 41, 94032 Passau, Germany
{stefan.becher,felix.boelz,armin.gerl}@uni-passau.de

Abstract. Privacy is an important topic regarding the modern and interconnected world. However, there is a tendency to ignore own intentions regarding customization of privacy policies in favor of saving time and using a specific service. To tackle this problem, a wide range of formal languages, expressing user preferences or privacy policies, have been developed by the scientific community. However, due to the variety of these languages, interoperability is only given between approaches designed to complement each other. As a result, as long as there is no standard in this domain, there will not be a broadly accepted solution. The aim of this paper is to provide a mechanism to counteract the lack of compatibility. To contribute to that very goal, this paper introduces the *Privacy Interfaces* mechanism which allows a *Unified Privacy Policy Negotiation*. It enables compatibility between otherwise incompatible preference languages and privacy languages, by creating standardized *Language Mappings*, which serve as a baseline for negotiation.

Keywords: Privacy · Privacy policy language · Privacy preferences · Unification · GDPR

1 Introduction

The awareness of individuals as well as governments about the importance of data privacy (especially on the internet) has steadily increased during the past years due to economical and governmental reasons. The processing of personal data is regulated more strictly in the European Union since the General Data Protection Regulation (GDPR) [9] was released in 2018. In response to GDPR, technical safeguards regarding personal data have to be adjusted. Traditional, textual privacy policies cause problems because often they cannot be comprehended without expert knowledge and it is impossible for users to read every policy they come across [15]. A better solution is to present the user a customizable privacy policy for a web service. There are mandatory options, which are required by the base functionality of the service as well as non-mandatory options which further enhance the user experience. For instance, data collection

S. Katsikas and S. Furnell (Eds.): TrustBus 2022, LNCS 13582, pp. 17–32, 2022.
https://doi.org/10.1007/978-3-031-17926-6_2

for user-tailored advertisements or product recommendations is a common non-mandatory option. In an ideal world, a user would need to enter his privacy preferences only once for similar services. In reality, this setup process must be repeated for each service. As this gets very time consuming, users might give up on picking non-mandatory options, especially for sensitive information, like health data. This leads to a worse experience for both user and service provider. It is important to deploy a fair agreement mechanism to satisfy both sides. Therefore, privacy languages and preference languages are proposed. Firstly, privacy languages model machine-readable, customizable representations of privacy policies. Secondly, preference languages express a set of user preferences regarding choices of privacy policy options. Instead of burdening the user with reading and understanding each privacy policy by his own these languages compare user preferences against privacy policy statements and summarize the results for the user. This process should function as automatically as possible and is in the following denoted as policy negotiation.

We show that there are numerous approaches for privacy languages and preference languages, whereas each approach sets a specific focus and comes with its own syntax and semantics. But so far no efforts have been made by research towards a holistic study regarding interoperability of these languages. Therefore, we state the following research question: *How can interoperability between existing approaches for privacy languages and preference languages with diverse syntax and semantics be granted?* After analyzing related work, we propose our solution for the interoperability problem. Thus, the main contribution of this work is the development of an instance between preference and privacy languages, called *Privacy Interfaces*, to resolve the aforementioned research question. *Privacy Interfaces* make it possible to automatically negotiate privacy policies of otherwise incompatible languages. To achieve this goal, privacy and preference languages are mapped to a unified privacy vocabulary registry. Registered languages share the same syntax and semantics and can therefore negotiate each other. This resolves the problem of missing compatibility and standards.

The remaining of the paper is structured as follows: In Sect. 2 related work is discussed regarding interoperability. In Sect. 3 a use case is presented which serves as a basis of this work. In Sect. 4 the *Privacy Interfaces* model is introduced. Section 5 evaluates the *Privacy Interfaces* principle. Finally, Sect. 6 discusses the results of this work and Sect. 7 concludes the work and gives an outlook for future research.

2   Related Work

In this section we discuss related work regarding privacy policy languages (PP), preference languages (P) and bundled approaches (PP, P) based on the compatibility with other approaches. Compatibility defines, if the languages were specifically designed to complement each other and an out-of-the-box-matching is possible. In addition, utilized vocabularies are compared, whereas some languages are based on their own vocabulary and others exploit generic

Table 1. Overview of privacy related languages, categorized into privacy policy language (PP) and preference language (P), used vocabulary (own vocabulary or generic for selectable external vocabulary), GDPR-compliance, compatibility with other languages and artifact format. We make no claim to completeness.

Language	Type	Vocabulary	GDPR	Compatibility	Artifacts
P3P [7]	PP	P3P	No	APPEL, XPref, SemPref	XML
P2U [11]	PP	Generic	No	No	XML
JACPol [12]	PP	Generic	No	No	JSON
SecPal4P [3]	PP, P	Generic	No	No	SecPal
EPAL [2]	PP	EPAL	No	No	XML
LPL [10]	PP	Generic	Yes	No	JSON, YAML, XML
SPECIAL [17]	PP, P	DPV	Yes	No	OWL2, RDF
PPL [18]	PP, P	PPL	Yes	XACML	XML
XACML [8]	PP	Generic	No	PPL	XML
CPL [13]	P	CPL	No	No	XML
APPEL [6]	P	P3P	No	P3P	XML
XPref [1]	P	P3P	No	P3P, APPEL	XML, XPath
SemPref [14]	P	P3P	No	P3P, APPEL	XML
YaPPL [19]	P	Generic	Yes	No	JSON, YAML

placeholder-vocabularies. Lastly, GDPR compliance is checked and the artifact types or outputs of the presented language are compared (see Table 1). In the following, each language is analyzed based on the mentioned criteria.

The privacy policy language P3P [7] introduces the concept of *sticky policies* to make sure that the data does not get disconnected from its original privacy policy. Moreover, a data retention approach is included. XACML [8] is an access control language to express and interchange privacy policies. As an extension to XACML, the Primelife Policy Language (PPL) [18] utilises the concept of data access and processing control. The language features so-called obligations, which give users the option to define rules, e.g., for receiving information on policy updates. P2U [11] sets a focus on secondary data usage by introducing price tags for data trading. The privacy policy language LPL [10] adds personal privacy and provenance control to privacy policies with a layered policy architecture. Personal privacy refers to personal privacy models which are stored for each data subject and define the degree of de-identification when data is used. The JACPoL [12] privacy language provides Attribute-based-Access-Control (ABAC) with efficient policy evaluation. SecPAL4P [3] features sets of assertions and queries to model both privacy policies and preferences in order to enforce access control. The syntax is based on SecPAL. As a more recent privacy language SPECIAL [17] introduces a privacy model which determines how data must be used and de-identified. While all these privacy languages share the main function to express privacy policies and usually provide a certain degree of customization for the data subjects, each of them is designed to fulfill a special purpose. Therefore, unique syntax elements and semantics are present in almost every privacy language.

The distinction between a privacy policy language and a preference language is in some cases fluent. SecPAL4P [3] already unites both concepts in one framework. In this case preferences are defined as SecPAL claims, which need to be fulfilled by the counterpart privacy policy statements. Also PPL [18] offers the base functionality of preferences as well as the option to define obligations. Other privacy languages are supported by one or multiple preference language. In the case of P3P a matching preference language is APPEL [6]. Other preference languages which work with P3P are XPref [1] which extends APPEL rules by a subset of XPath and SemPref [14] which adds semantics to APPEL rules. Finally, several preference language approaches exist which do not have a counterpart privacy language yet. One of them is YaPPL [19]. It considers GDPR requirements by modelling explicit consent and is designed to be a lightweight preference language for IoT applications. CPL [13] is a preference language with a special focus on context information. Users can express preferences which consider time and location constraints of data processing.

Most of the discussed languages use their own custom privacy vocabulary or are defined generically so that the vocabulary can be chosen based on the application. For example, P3P comes with its own P3P vocabulary [7], which includes a predefined set of purposes, recipients, data categories and access methods. However, this diversity in syntax and semantics works against interoperability. To enable a matching between multiple different privacy and preference languages, it is necessary to agree on a common format and a baseline vocabulary.

The most recent efforts in creating a standardized vocabulary for privacy handling is done by the W3C Data Privacy Vocabularies and Controls Community Group (DPVCG) which develops the Data Privacy Vocabulary (DPV) [5,16] as successor of the SPECIAL vocabulary [4]. This vocabulary is centered around personal data handling and is defined as an RDF graph.[1] The graph is extendable with custom terms which is an advantage in terms of standardization and interoperability. Also, DPV has a GDPR-tailored extension, which provides a legal basis, that is mandatory for European privacy policies [9]. We see this vocabulary to be the most promising candidate to reach a broad application as it covers recent changes for legal privacy handling and its development is supported by industry.

3   Running Example

To give a better understanding of the *Privacy Interfaces* principle, we will refer to the following use case during the course of this paper. Lets assume an online doctor's practice, where patients can call doctors via webcam and receive remote diagnosis. For the usage of this service the patients must provide several personal and sensitive information. An extract of the privacy policy of the practice might look like the following: *"Collected personal information, held at the*

---

[1] The namespace of the DPV is http://www.w3.org/ns/dpv#.

practice as electronic records will include patient's contact information including name, address and phone number and medical information including medical history, medications and risk factors. Personal information may be used or disclosed for medical purposes including diagnosis and medical treatment of the patient or optional advertisement including monthly shipment of health magazine by advertising partners". This exemplary privacy policy will be expressed in a privacy language to achieve machine-readability. For this use case we have chosen the Layered Privacy Language (LPL) because of its support of GDPR and JSON-policies, which are human-readable. An extract of the converted LPL privacy policy might look like the following example shown in Listing 1.1.

```
1  "layeredPrivacyPolicy": {
2    "purposeList": {
3      "purpose": [
4        {"name": "Medical purposes", "required": true,
5          "dataRecipientList": {
6            "dataRecipient": [{
7              "name": "Doctors practice", "thirdCountryTransfer
                 ": false }] },
8            "dataList": {
9              "data": [{ "name": "Medical information" }] } },
10        {"name": "Advertising", "required": false,
11          "dataRecipientList": {
12            "dataRecipient": [{
13              "name": "Magazine agency", "thirdCountryTransfer
                 ": false }] },
14            "dataList": {
15              "data": [{ "name": "Contact information" }] } }] }
                 }
```

Listing 1.1. Privacy policy of an online docter's practice displayed by LPL

```
1  {"preference" : [{
2    "rule" : {
3      "purpose" : {
4        "permitted" : ["Medical treatment"],
5        "excluded" : ["Advertising"] },
6      "utilizer" : {
7        "permitted" : ["Online Doctors Practice"],
8        "excluded" : ["Advertising Partner"] } } }] }
```

Listing 1.2. YaPPL preference for a user who wants to share contact data and medical data for medical treatment and administrative tasks.

Next, we assume a user who wants to get online diagnosis because of respiratory problems. The user is unsure which service he wants to use because he is concerned about possible leaks of his sensitive health data. He decides to make a preference setup in advance in order to safe time when negotiating the privacy policy. The user decides to share his contact and medical data medical purposes with the online doctors practice while rejecting advertisement. A possible extract of the preference modelled by the YaPPL preference language is shown in Listing 1.2. We have chosen YaPPL because it also considers GDPR and the preferences are defined in JSON format, too. Note, that is it not possible to restrict the usage of data categories within YaPPL preferences. As we can see, there are differences in the syntax and the semantics of both languages. The structure of LPL is more hierarchical than the structure of the preference. Each LPL purpose element contains data recipients and data elements while for the preference all of this elements are on the same level. In addition, there are differences in the naming of the used vocabularies, e.g., *Medical purposes* and *Medical treatment* or *Doctors Practice* and *Online Doctors Practice*. Finally, there are elements which are unique to the languages, like the representation of the data categories contact and medical information of the policy or the excluded purpose *Advertising* of the preference. Therefore, an out-of-the box-matching is not possible. We propose a system to overcome this barrier by creating standardized representations of privacy languages and preference languages in the following.

4   Privacy Interfaces

In this section a generic *Language Mapping* process, called the *Privacy Interfaces*, is introduced. With this process, we aim to reach the goal of a high compatibility between privacy policy languages and preference languages. Firstly, both languages must be mapped on a standardized vocabulary in order to achieve unified semantics as a basis for negotiation (see Fig. 1). For this work we have chosen the Data Privacy Vocabulary because it arose from the most recent efforts to create a standardized vocabulary for privacy handling and takes GDPR into account. Those *Language Mappings* must be provided by a domain expert like for example a data protection officer, developer of the given language or similar. Secondly, if both mappings are given, an instance of the privacy policy language, can be automatically mapped on an instance of the preference language. We assume that the target privacy language does not utilize or extend the DPV. Otherwise, a translation into DPV can be realized more easily with Semantic Web technologies and the RDF graph. Thus, a baseline for all remaining privacy policy languages must be found to facilitate a constructive mapping. Therefore, it is key to take a look at the languages artifacts or outputs for a given privacy policy. Those outputs are mainly XML, JSON, YAML, or RDF files (see Table 1) while there can also be other formats. Unified negotiation is only possible if all outputs are converted to one of the formats or are abstracted to another representation.

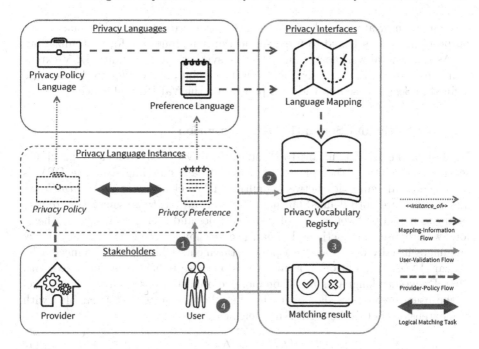

Fig. 1. Workflow model of the *Privacy Interfaces*. Non-compatible privacy and preference languages are mapped to a privacy vocabulary registry which manages matching.

4.1 Workflow

In the following, the workflow of the *Privcy Interfaces* is described (see Fig. 1).

The *Mapping-Information Flow* requires that some stakeholder provides a *mapping* defining how an input language gets translated into DPV. More precisely, a mapping refers to one or multiple DPV instances which, in combination, represent a privacy policy or user preference. The authority who provides the mapping might be a domain expert, like data protection officer, language developer or similar. The mapping can be defined by applying the *Language Mapping* via REST requests to the registry server. As soon as the mappings of both given languages is provided, the privacy vocabulary registry contains all necessary information to translate instances of both languages into DPV instances.

The *User-Validation Flow* refers to the users' point of view. As a first step, the user provides the privacy vocabulary registry with his or her privacy preference. In addition to that, a privacy policy must be provided by an engaged service, or forwarded by the user. At the second step, the registry translates the instances with the help of the previously provided mappings into the DPV. During the third step, both instances are automatically matched and the policy is negotiated, by utilizing the full functionality of the chosen preference language. Finally, the user is presented the result of the matching. The presentation of the result can include the matching status of mandatory purposes as well as recommendation or automated negotiation of optional purposes. While mandatory purposes must be selected to use a service, optional ones are for customization.

As the general workflow is outlined, the syntactic and semantic DPV extension is detailed in the next section. From this point on, we refer to the use case defined in Sect. 3 based on the LPL policy and the YaPPL preference.

4.2 Syntactic and Semantic DPV Extension

By design, the DPV is meant to standardize privacy expressions by providing several hierarchies in the context of personal data handling. But the DPV is missing logical semantics between multiple DPV elements, e.g. *MedicalHealth* and *Contact* are both defined but not related. However, these semantics are necessary to model preference- or policy-based rules and their relationships. Therefore, we propose a syntactic and semantic extension for the DPV. Privacy policy languages usually define a list of policy statements which are all connected by logical *ANDs*. In addition, optional parts of those policies are appended with *ORs*. Preference languages might be more complicated based on the detail level of the users preferences. For example, the following statement is enabled with this extension. Let E_i be a DPV-root element.

$$(E_1) \vee (E_1 \wedge E_2) \tag{1}$$

In any case E_1 must be valid, whereas in addition the optional element E_2 could be valid, too. This equation models the previously given privacy policy (refer to Listing 1.1) with the purposes *Medical purposes* and *Advertising* as E_1 and E_2 respectively. For this equation it is assumed that the respective data categories and recipients are correctly included in the DPV-root elements. It is also possible to set the DPV sub-root elements into relation. The data *Medical information* and *Contact information* are represented by D_1 and D_2, the purposes *Medical purposes* and *Advertising* are modeled by P_1 and P_2 and the data recipients *Doctors practice* and *Magazine agency* refer to R_1 and R_2. The following equation models their relationship as given in the LPL policy.

$$(P_1 \wedge R_1 \wedge D_1) \vee ((P_1 \wedge R_1 \wedge D_1) \wedge (P_2 \wedge R_2 \wedge D_2)) \tag{2}$$

The mapping of the previously defined preference is similar. For the abbreviations refer the previous equations. As the preference might have a different vocabulary than the policy, elements with a different name might refer to the same entity. The recipients *Online Doctors Practice* and *Advertising Partner* refer to R_1 and R_2. The equation in the following models this preference.

$$(\neg P_2) \wedge (\neg R_2) \wedge R_1 \wedge (P_1 \vee P_2) \tag{3}$$

Therefore, the elements of the privacy policy and the preference are mapped to the elements of the DPV extension and can be logically matched. For the given use case the mapping will give a positive outcome for (E_1) because *Medical purposes* are permitted but for $(E_1 \wedge E_2)$ the outcome will be negative because

Advertising is rejected. To unify the syntax of the DPV extension we propose the following notation (see Fig. 2). Core of the syntax is a single DPV element, which represents any entry of a DPV hierarchy. An optional unary operator can be defined before the DPV element and is mainly used to negate it. Several DPV elements can be logically joined together by binary operators. Supported operators are *NOT*, *AND*, *OR* and *XOR*. As the syntactic and semantic DPV extension is given, the *Language Mapping* process is explained next.

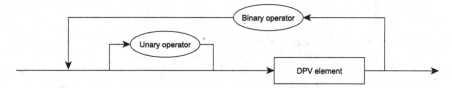

Fig. 2. Syntax diagram of the DPV extension. An optional *unary operator* is followed by a *DPV element*. *DPV elements* are logically connected by *binary operators*.

4.3   Privacy Language Mapping

For the scope of this paper, the mapping is only realized for JSON files because it is resource efficient, human-readable and commonly used in other domains, like the Semantic Web and IoT. The goal is to provide mechanisms to firstly access data in a JSON file, and to secondly describe its relations to the DPV. Those mechanisms combined describe how to translate any privacy policy of a given privacy language or dialect of it, which is in the following called *Language Mapping* or *Privacy Interfaces*. Dialect refers to the possibility, that services use the same privacy language but different vocabularies. Therefore, the usage of locally defined languages can become globally valid by mapping the dialects.

The *Privacy Interfaces* include a set of *Language Mappings* (see Fig. 3). Every *Language Mapping* defines a set of *Complex Objects*. These define how the DPV root component, the *PersonalDataHandling*, shall be constructed based on the given privacy language. *Complex Objects* might be related to any *Constructable Object*. Those concepts are described in more detail in the following. The *Privacy Interface* model supports extensibility because it is sufficient to implement a new *Constuctable Object* and provide its REST interface for any extension.

The most basic object, *Primitive*, defines any primitive value, e.g., *String*, *Integer*, *Boolean*, etc., which is allowed by the JSON format. *Action* is used to process primitive functions on any *Constructable Object*. It defines the operators *Plus* ($+$), *Minus* ($-$), *Divide* ($\div$), and *Multiply* ($*$). The *RootedRessource* extracts data from a given JSON privacy policy. Those files consist of recursive key-value pairs. Therefore, any primitive JSON type or object can be referred to as a path of keys. JSON arrays can also be accessed by a path whereas it is possible to additionally define whether all of its fields shall be extracted or only defined indices. The *Condition* allows to construct an if-else condition, while the *Loop* is used to traverse and modify arrays. The objects *Duplicate* and *ExceptionHandler* are used for technicalities and error handling.

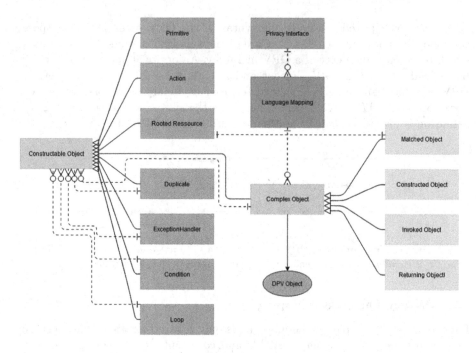

Fig. 3. Structure of the *Privacy Interfaces*. Each *Language Mapping* consist of a set of *Complex Objects* which define how the DPV root element is constructed. *Constructable Objects* define how data is accessed from a file and how to map it to DPV (sub-)classes.

The four *Complex Objects*, allow the usage of constructors, getter and set-ter methods, and additional helper methods, for instance to modify and format dates. The *Constructed Object* is able to invoke any constructor. The *Invoked Object* is similar to the *Constructed Object* but invokes methods instead of constructors. However, this component does not return the result of the method, but the object after given method was invoked on it. The *Returning Object* is a subtype of the *Invoked Object*. This object returns the result of the invoked method. The *Matched Object* translates a given String into its respective DPV category. If an extraction is not possible, a given restriction will be chosen, e.g. a default super-category. As the *Privacy Interfaces* and their workflow are now given, a quantitative and qualitative evaluation is conducted to review the effectiveness of the proposed model.

5   Evaluation

All components of the workflow have been implemented in a Java prototype framework, which shows on the one hand a proof of concept, and on the other hand the feasibility regarding the computational effort. In order to conduct the performance evaluation, a qualitative evaluation must be done first, which shows the applicability of the *Privacy Interfaces*. In the following, a DPV-mapping for the exemplary LPL privacy policy shown in Sect. 3 is created and evaluated.

5.1 Mapping Evaluation

As the process of creating the *Privacy Interfaces* is relatively free, the LPL mapping could be designed in multiple ways. The mapping performance also depends on this design. The task is to develop a valid *Language Mapping* to be able to translate LPL policies into DPV instances. For each DPV element, the conceptual mapping of LPL is given, starting with *Purpose*. In contrast to the DPV, LPL *Purposes* include components describing the *Data* or the *DataRecipient*. In the DPV schema however, those concepts are top level elements and therefore will be mapped separately. For now, the purpose itself is the only relevant part. The mapping is created as follows (see Listing 1.3).

First of all, the DPV root element *PersonalDataHandling* is initiated with a *ConstructedObject*. This is referred as step 0. This element represents the policy mapping and will be extended by each of the following mappings. Each LPL purpose consists of a *name* and a *required* field, which describes whether it is optional or not. The *optional* behavior is supported by the DPV extension by connecting all DPV purposes with logical expressions as shown in Sect. 4.2. Both fields are extracted by a *Rooted Resource*, shown by steps 1 and 2 in the listing. Step 3 is realized with the *Matching Object*, which initializes the DPV *Purpose* mapping by its name or type, and its required value. Finally, with step 4, the constructed *Purpose* mapping is appended to the *PersonalDataHandling* mapping with an *InvokedObject*.

Listing 1.3. The LPL-DPV mapping considering the purposes only

```
1 /** 0. Construct the DPV element */
2 ComplexObject<PersonalDataHandling> policy = new
    ConstructedObject()
3
4 /** 1. Resolve the LPL Purpose names */
5 RootedResource names = new RootedResource("
    layeredPrivacyPolicy/purposeList/purpose/name",String.
    class)
6
7 /** 2. Resolves if the purposes are required */
8 RootedResource required = new RootedResource("
    layeredPrivacyPolicy/purposeList/purpose/required",
    Boolean.class)
9
10 /** 3. Build Purposes based on their name */
11 ComplexObject<Purpose> purpose = new MatchingObject(names,
    Purpose.class,required)
12
13 /** 4. Add Purposes */
14 policy = new InvokedObject(policy,"setPurpose",purpose)
```

Steps 1 to 4 (extraction, creation, appending) are repeated for every DPV sub-element. The only differences are the locations within the LPL structure, and slightly different construction schemes for the other DPV components.

For instance, a DPV *DataSubject* requires its name and potentially other personal information as inputs, but not a required tag as for *Purpose*. Additionally, for technical issues other *Privacy Interface* components will be used to handle errors, null values, formatting, if-else cases or similar. For instance, the mapping of all LPL information related to the DPV *Processing* is a bit more complicated. In particular, the LPL field *thirdCountryTransfer* describes the type of the processing. Therefore, after the extraction via a *Rooted Resource*, a *Condition* injects the DPV *Transfer* with a *Constructed Object* as a DPV *Process*. The condition will return the *Transfer* if the *thirdCountryTransfer* is *true*, otherwise this *Process* will not be added. Similarly, this process will be independently applied to the DPV *Transfom* category if, for instance, an anonymization or a pseudonymization is given. As the LPL privacy policy mapping is now given, it can be used to translate any LPL policy into DPV instances. Such *Language Mappings* can be created for any privacy language or preference language by mapping the corresponding DPV elements.

5.2   Processing Performance

In order to show the feasibility of the previously proposed approach, a processing performance evaluation[2] based on the *Language Mapping* created before is given in the following. The measurements assume that the mapping of the privacy policy language is stored in the privacy vocabulary registry. The measured mapping process includes the parsing of the JSON file (the privacy policy instance) into internal Java objects. Explicitly excluded are the conversion of preference language objects into the DPV schema (the user preference instance) and the policy negotiation process. Also, it is important to mention that the complexity mainly depends on the provided mapping mechanism. As the mechanism enables arbitrary nested constructors, methods and loops, the mapping can also be designed in an inefficient way. In the following, the performance of the translation step (LPL policy into DPV instance) on different sized policies is measured. All input LPL instances have a similar structure to the presented use case example, but include all of LPLs defined data fields. Figure 4 shows the resulting DPV root elements in relation to the necessary conversion time in ms. The dashed line represent a linear growth, the dotted one a polynomial one. In this evaluation, the polynomial trend is relatively flat.

In order to answer the question, whether the given approach is applicable in real-time, it highly depends on the size of the input privacy policy. The linear and super-linear trend line would reach the limit of half a second for a file with 84 DPV instances. The implementation of caching mechanisms and DPV extensions can possibly further improve the performance. As a result, real use cases can be achieved in real-time, namely in linear time complexity. Thus, the given approach scales for realistically sized privacy policies.

---

[2] The Performance is measured on a 64-bit Microsoft Windows 10 system with 64 GB DDR4 RAM, an Intel Core i7-8750H CPU @2.20 GHz with 12 logical processors, on Java 15 with a maximum of 18 GB Heap size.

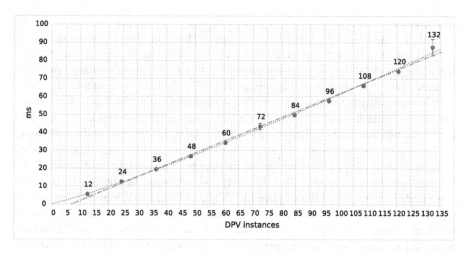

Fig. 4. Required time in ms for translation by size of the LPL policies (measured in number of created DPV root instances). Linear to slightly polynomial complexity.

6   Discussion

In this work we showed the lack of interoperability in current state-of-the-art approaches for privacy languages and preference languages. Usually, each preference language is designed to match exactly one privacy policy language. This is mainly caused by missing standardization shared between the approaches. On the one hand, there are syntactical incompatibilities whereas translation is needed between different language structures in order to achieve a common basis for matching. On the other hand, there are semantic inconsistencies due to the lack of wording standards.

As a solution to increase interoperability, we propose the principle of *Privacy Interfaces*, which serve as a negotiation middleware between otherwise incompatible languages. Existing privacy policy languages and preference languages are not altered by our approach, meaning that users and providers can freely choose. Instead, we provide a generic structure to create *Language Mappings* for both privacy policy languages and preference languages on the basis of the DPV privacy vocabulary. As the DPV lacks semantic connections between elements of several data handling categories we created a DPV extension which is able to express real policy statements. The *Language Mappings* are registered in a privacy vocabulary registry and can be accessed during privacy policy negotiation. Therefore, unified preference matching for privacy policies is enabled. For the evaluation of our approach we did a full *Language Mapping* of the Layered Privacy Language (LPL). The run time of the complete mapping process of the LPL policy shows a slightly super-linear growth by an increasing amount of DPV instances. As we evaluated the mapping of the whole language it is a reasonable assumption that this process is conductable in real time for any privacy policy language.

Limitations. In theory, all privacy languages and preference languages can be integrated in the privacy vocabulary registry. However, there are some limitations which are derived from the DPV. If the DPV, including the introduced extension, is not able to map a language construct of a given privacy language, it cannot be integrated in the language registry. To fix this problem the DPV must be extended by new elements. These elements might originate from other privacy vocabularies and can contribute to a more complete version of the DPV. As long as new elements are only added to the DPV hierarchies and no elements are removed or their ordering within the RDF graph is changed, compatibility is still given for previously created mappings. The current prototype of the *Privacy Interfaces* only supports JSON-typed artifacts. In the future this will be extended for other artifact types like XML or YAML. While it is an advantage, that mappings can be formulated nearly arbitrarily, due to the very generic structure of the mapping interface, it also requires expert knowledge. Therefore, the initial integration of a new privacy policy language or preference language might be done by a privacy officer, language developer or other domain expert.

7   Conclusion

With our work we showed that standardization in the field of privacy enhancing technologies is not only possible but also needed. We created an overarching framework for interoperability in order to unify privacy handling technologies. We introduced the principle of so called *Privacy Interfaces* which enable *Language Mappings* for almost any privacy language and preference language to a standardized privacy vocabulary. These *Language Mappings* are based on the Data Privacy Vocabulary, which we extended by semantics to be able to express regular policy statements. The unified statements are later used for privacy language independent preference matching.

Based on the use case of an online doctors' practice we show that an out-of-the-box-matching of incompatible languages is not possible and how this problem is solved by the introduction of the *Privacy Interfaces*. Therefore, we conduct a full *Language Mapping* of the Layered Privacy Language and evaluate the run-time performance of the mapping process. The results show that the run-time in the range of milliseconds grows super-linearly by the amount of created DPV root elements, representing an element of the privacy policy. As the mapping was performed on realistically sized privacy policies and no exponential growth occurred, the mapping of a complete language is possible in real-time.

For future work, we envision to further enhance the current *Privacy Interfaces* prototype. For an easy and secure integration of *Language Mappings* we will implement and evaluate a user interface to replace the current REST interface. Furthermore, we are working on an algorithm to verify the correctness of *Language Mappings*. To extend the applicability of privacy preferences we finally plan to also create *Language Mappings* for plain-text privacy policies. Therefore, we will analyse policies by machine-learning techniques and automatically label their content with expressions of our DPV extension.

As new legal regulations, like the GDPR, and public concern about privacy emerge, it is more important than ever to support this process with secure and sustainable technologies. These technologies can only be successful, if they are widely accepted and cover a broad amount of applications. We hope to motivate further research towards standardization in the field of privacy protection to realize this goal in the future.

References

1. Agrawal, R., Kiernan, J., Srikant, R., Xu, Y.: XPref: a preference language for P3P. Comput. Netw. **48**(5), 809–827 (2005)
2. Ashley, P., Hada, S., Karjoth, G., Powers, C., Schunter, M.: Enterprice privacy authorization language (EPAL 1.2), Tech. Rep., IBM (2003)
3. Becker, M.Y., Malkis, A., Bussard, L.: A framework for privacy preferences and data-handling policies. Microsoft Research Cambridge Technical Report, MSR-TR-2009-128 (2009)
4. Bonatti, P., Kirrane, S., Petrova, I., Sauro, L., Schlehahn, E.: The special usage policy language v1.1 (draft), Tech. Rep., SPECIAL H2020 (2022). https://ai.wu.ac.at/policies/policylanguage/
5. Bos, B., et al.: DPV vocabulary specification (2022). https://www.w3.org/ns/dpv-gdpr
6. Cranor, L., Langheinrich, M., Marchiori, M.: A P3P preference exchange language 1.0 (appel 1.0): w3c working draft 15 April 2002. World Wide Web Consortium (W3C) (2002). https://www.w3.org/TR/P3P-preferences/
7. Cranor, L., Langheinrich, M., Marchiori, M., Presler-Marshall, M., Reagle, J.: W3C P3P specification (2002). https://www.w3.org/TR/P3P/
8. Rissanen, E., Bill Parducci, H.L.: eXtensible access control markup language (XACML) version 3.0. Tech. rep., OASIS (2013). http://docs.oasis-open.org/xacml/3.0/xacml-3.0-core-spec-os-en.html
9. European-Commission: Regulation (eu) 2016/679 of the European parliament and of the council on the protection of natural persons with regard to the processing of personal data and on the free movement of such data (general data protection regulation). https://eur-lex.europa.eu/eli/reg/2016/679/oj
10. Gerl, A., Bennani, N., Kosch, H., Brunie, L.: LPL, towards a GDPR-compliant privacy language: formal definition and usage. In: Hameurlain, A., Wagner, R. (eds.) Transactions on Large-Scale Data- and Knowledge-Centered Systems XXXVII. LNCS, vol. 10940, pp. 41–80. Springer, Heidelberg (2018). https://doi.org/10.1007/978-3-662-57932-9_2
11. Iyilade, J., Vassileva, J.: P2U: a privacy policy specification language for secondary data sharing and usage. In: Proceedings IEEE Security and Privacy Workshops, pp. 18–22 (2014)
12. Jiang, H., Bouabdallah, A.: JACPoL: a simple but expressive JSON-based access control policy language. In: Hancke, G.P., Damiani, E. (eds.) WISTP 2017. LNCS, vol. 10741, pp. 56–72. Springer, Cham (2018). https://doi.org/10.1007/978-3-319-93524-9_4
13. Kapitsaki, G.M.: Reflecting user privacy preferences in context-aware web services. In: IEEE 20th International Conference on Web Services, pp. 123–130 (2013)
14. Li, N., Yu, T., Antón, A.: A semantics-base approach to privacy languages. Comput. Syst. Sci. Eng. - CSSE 21 (2006)

15. McDonald, A.M., Reeder, R.W., Kelley, P.G., Cranor, L.F.: A comparative study of online privacy policies and formats. In: Goldberg, I., Atallah, M.J. (eds.) PETS 2009. LNCS, vol. 5672, pp. 37–55. Springer, Heidelberg (2009). https://doi.org/10.1007/978-3-642-03168-7_3

16. Pandit, H.J., et al.: Creating a vocabulary for data privacy. In: Panetto, H., Debruyne, C., Hepp, M., Lewis, D., Ardagna, C.A., Meersman, R. (eds.) OTM 2019. LNCS, vol. 11877, pp. 714–730. Springer, Cham (2019). https://doi.org/10.1007/978-3-030-33246-4_44

17. SPECIAL-H2020: Scalable policy-aware linked data architecture for privacy, transparency and compliance. https://specialprivacy.ercim.eu/

18. Trabelsi, S., Njeh, A., Bussard, L., Neven, G.: PPL engine: a symmetric architecture for privacy policy handling. In: W3C Workshop on Privacy and Data Usage Control, vol. 4 (2010)

19. Ulbricht, M.-R., Pallas, F.: YaPPL - a lightweight privacy preference language for legally sufficient and automated consent provision in IoT scenarios. In: Garcia-Alfaro, J., Herrera-Joancomartí, J., Livraga, G., Rios, R. (eds.) DPM/CBT -2018. LNCS, vol. 11025, pp. 329–344. Springer, Cham (2018). https://doi.org/10.1007/978-3-030-00305-0_23

A Constructive Approach for Raising Information Privacy Competences: The Case of Escape Room Games

Thanos Papaioannou[1]([✉]) [iD], Aggeliki Tsohou[1] [iD], Georgios Bounias[1] [iD], and Stylianos Karagiannis[2] [iD]

[1] Ionian University, Corfu, Greece
{thanospapa,atsohou}@ionio.gr
[2] PDM & FC, Lisbon, Portugal
stylianos.karagiannis@pdmfc.com

Abstract. Creating educational interventions to enhance information privacy awareness is at the heart of research, as the protection of personal data and privacy is increasingly of concern to internet users. However, although efforts have interactive and exploratory features, they are not sufficiently supported by a theoretical learning framework, such as constructivism. In addition, they do not target specific learning outcomes to prepare competent users to protect their own privacy and personal data. Based on the requirements of the Constructivist Information Privacy Pedagogy and the Information Privacy Competency Model for Citizens we claim that escape rooms are the ideal constructive educational interventions, and we propose an indicative scenario for an escape room oriented towards specific competences for the internet user.

Keywords: Constructivism · Privacy competences · Serious games · Escape rooms

1 Introduction

As the daily life of people is connected to, and dependent on, the use of the internet, the threats against information security and privacy of internet users continue to increase in number and variety. Regulatory interventions around the world, such as the General Data Protection Regulation of the European Union, highlight the importance of information security and privacy. Further, the implications of information privacy issues that often arise and concern users, highlight information privacy as one of the most important ethical issues of our time (Lavranou and Tsohou 2019).

Several studies indicate the need for systematic educational support for internet users, through all forms of education, formal and non-formal, in order to acquire the knowledge and skills to meet the challenges posed by privacy issues (Lavranou and Tsohou 2019). The knowledge and skills that the internet user should have to meet the above challenges, such as of the controlled disclosure of personal information or the configuration of privacy settings, have become a subject of relevant research, highlighting the attributes of privacy competent internet users (Tsohou 2021). The above efforts target the development of a privacy competency model, on which educational programs oriented can rely.

To date, the term privacy awareness is the main one used to describe the knowledgeable and competent internet user. At the same time though, privacy awareness is a general term defined as individuals' understanding of concepts related to the protection of personal data and privacy, including privacy concerns and privacy risk perceptions (Soumelidou and Tsohou 2019; Tsohou 2021). Some efforts to raise users' privacy awareness are based on presenting privacy policies in different forms (i.e., visual instead of textual). Other privacy raising techniques include software tools, such as transparency enhancing ones (Malandrino et al. 2013) to inform the user about personal data that are shared. Other educational interventions, include innovative methods, such as gamification, game-based learning, etc. (Soumelidou and Tsohou 2019). However, these efforts are not based on a theoretical foundation that can adequately support the design of the educational interventions.

A common approach in educational sciences to supporting learning interventions is to relate them to existing theories, which explain the phenomenon of human learning. Especially in interventions related to playful or exploratory learning, researchers invoke the learning theory of constructivism rather than the older "traditional" theories, such as behaviorism and cognitivism. According to constructivism, the individual constructs her knowledge from her experiences and interactions with the environment and other people and is not a passive recipient of information. Constructivism is considered the most modern learning theory and the educational interventions based on it the most effective (Bada and Olusegun 2015). Although most educational interventions invoke their constructive dimension, they are not sufficiently supported by the relevant learning theory (Papaioannou et al. 2022). The main reason for this deficit is, until recently, the lack of a constructive pedagogy from the literature (Richardson 2003), and especially a pedagogy focused on information privacy.

To date, several proposals for educational interventions on privacy have been presented, either with constructive and playful features, or not. The type of these interventions ranges from software tools that monitor the flow of personal data (EnckWilliam et al. 2014; Hatamian et al. 2017), applications for providing personalized advice on good privacy practices (Ghazinour et al. 2016) and interactive software tools that enhance user's awareness (Aktypi et al. 2017; Tadic et al. 2018) to playful activities aimed at mobilizing the player and informing her about information privacy and personal data protection (Cetto et al. 2014; Pape 2022; Raynes-Goldie and Allen 2014; Suknot et al. 2014). Researchers argue that traditional privacy awareness interventions, such as lectures and presentations on privacy policies, are not always effective in practice (Lavranou and Tsohou 2019; Soumelidou and Tsohou 2021). In addition, existing attempts of modern educational approaches do not have a more concrete educational goal towards information privacy, such as the cultivation of specific competences of the internet user and they are not adequately based on a learning theory. Thus, the following research questions arise: "What form should an educational approach take to increase competences for privacy protection?" and "How can educational approaches increase privacy competences?". In this paper we examine the relevant literature on privacy competences and on privacy educational methods. Through this analysis we reveal that privacy education attempts in the literature lack theoretical foundation and the definition of concrete educational goals. To address this gap, we suggest an educational intervention that comprises a constructive learning theory and game-based learning. As an exemplary application of this framework, we suggest the utilization of escape rooms for the realization of the game-based learning aspects.

The paper is structured in six sections. Following this introduction, Sect. 2 discusses existing work and theoretical framework on information privacy awareness and competences. Section 3 presents Constructivism learning theory in the design of educational interventions. In Sect. 4 our proposed framework for raising Information Privacy Competences through the exploitation of educational escape rooms is analyzed, followed by an indicative case scenario in Sect. 5. Finally, Sect. 6 concludes the paper, discussing implications and future work.

2 Information Privacy Awareness and Competences

2.1 Information Privacy Awareness

Researchers agree that awareness is a key motivator for a person to protect their personal data and privacy (Gerber et al. 2017). There is no common definition of privacy awareness (Soumelidou and Tsohou 2019). Definitions given concern users' control over the people who check their social media profiles and to user's concerns about their privacy, individuals' knowledge of the potential threats that may arise from the disclosure of her personal data, the consequences that they may bring, the knowledge of privacy policies and terms of use, as well as inferences that one can produce from social media information about someone (Soumelidou and Tsohou 2019). Combining the above considerations, Sim et al. (2012) developed a holistic definition that extends from the individual's knowledge of privacy policies to their privacy concerns.

The importance of privacy awareness is directly associated with preventing and addressing privacy threats, such as illegal monitoring and interception of economic interest information (Joinson et al. 2010). Users' role is crucial for threat prevention, particularly by paying attention to the personal data that they disclose (Soumelidou and Tsohou 2019). Nevertheless, research findings suggest that users are not fully aware of the privacy issues that may arise from their online behavior (Soumelidou and Tsohou 2019). Other researchers highlight the importance of users' lack of awareness regarding processes that "run in the background" and are associated with personal data processing (Barth et al. 2019).

Consequently, researchers have investigated ways to increase privacy awareness. According to Soumelidou and Tsohou (2019) several approaches for privacy awareness raising are identified: (a) informing users for the potential threats when they disclose large amount of personal data (Malandrino and di Salerno 2012) and (b) providing personalized privacy recommendations (Cetto et al. 2014) through tools, extensions, frameworks, applications and other technologies. In sequence, we briefly indicate studies for each of the aforementioned approaches and means of privacy awareness enhancing.

Aktypi et al. (2017) created an interactive software tool which illustrates online information shared by wearable devices' users and elaborates on how they might be susceptible to unwanted leakage of their personal data and a privacy breach. The tool also highlights the privacy risks that appear from personal data that people disclose. Hatamian et al. (2017) have built a smartphone monitoring application which collects information regarding each installed application's access to sensitive resources in order to inform the user about the privacy invasiveness level of the mobile applications. A similar effort was implemented by EnckWilliam et al. (2014) with *TaintDroid* which is a taint tracking and analysis system that monitors the behavior of smartphone's installed applications, with regards to user's personal information misuse. Paspatis et al. (2020) have introduced *AppAware*, an innovative method of visualizing the permissions asked by applications installed on a smartphone, in order to make the user more interested in the way in which third parties access her personal data. Malandrino et al. (2013) introduced *NoTrace*, a tool that informs the user about the exact personal information that is shared with third parties and advertising companies. In this way the user becomes aware and makes better privacy decisions. Tadic et al. (2018) proposed the *CyberActivist* tool in order to help activists to understand and moderate privacy risks in such environments. Ghazinour et al. (2016) introduced *YourPrivacyProtector*, a software tool which uses machine learning techniques in order to classify SNS users into groups according to their privacy awareness level. Additionally, the tool provides relevant personalized recommendations to the user regarding her privacy protection.

A significant means utilized nowadays for increasing users' privacy awareness is privacy policies. Online service providers include in privacy policies and statements the information related to personal data processing. Many researchers agree on the role of privacy policies towards increasing users' privacy awareness (Soumelidou and Tsohou 2019). Despite their effectiveness, many studies provide evidence that users commonly do not read privacy policies (Acquisti and Gross 2006; Tsohou and Kosta 2017). Factors that determine whether users will read them are associated with their language, complexity and comprehensiveness, since users commonly state that they are "complex legal texts", difficult to read and understand (Tsohou and Kosta 2017). To address these deficits researchers have proposed to simplify privacy policies through visualization (Soumelidou and Tsohou 2019; Vemou et al. 2014) but this has not been yet realized by online service providers.

All the aforementioned research and practical efforts to increase users' privacy aware-ness lack two important, in our view, characteristics: (a) they do not have a focus on what specific learning outcomes they target, such as identified target competences for the internet user, and (b) their development is not based on theoretical foundation related to the creation of educational interventions. Thus, in this paper we aim to address this research gap.

2.2 Information Privacy Competences

Competencies refer to the characteristics of an individual that lead to superior perfor-mance (Winterton 2009). Competency commonly comprises three elements: knowledge, skills, and abilities/attitudes (Holtkamp 2015) that are necessary to solve a problem in a given context. Although, information systems researchers have studied privacy awareness, there is limited works concerning information privacy competences.

Tsohou (2021) proposed an Information Privacy Competency Model (IPCM), which highlights the attributes that internet users should hold in order to be competent to protect their personal data and privacy. The aggregated competencies can help the design and development of educational interventions towards improving users' privacy protective behavior. For the development of the competency model, Tsohou (2021) utilized the Iceberg model (Spencer and Spencer 2008) which classifies competences into knowledge, skills, attitudes, social role, traits, self-image, and motives. The privacy competencies identified are presented in Table 1.. Each competency has been assigned a reference number used hereafter in the paper.

Table 1. An information privacy competency model for citizens (Tsohou 2021)

Element	Competences
Knowledge	Knowledge about own self-disclosure (K1), Knowledge of safeguards (K2), Knowledge of privacy risks (K3), Knowledge of own personal data rights (K4), Knowledge of regulation and legislation (K5)
Skills	Metacognitive accuracy (S1), Ability to install and customize safeguards (S2), Ability to read and understand privacy policies (S3), Ability to understand online tracking (S4), Ability to understand technological environments (S5), Ability to exercise own rights (S6), Ability to perceive risks (S7)
Social role	Privacy concerns (SR1), Privacy (SR2), Self-control on personal data disclosure (SR3), Confidence (SR4), Freedom (SR5), Anonymity (SR6), Self-control (SR7), Personal Dignity (SR8), Fear-free living (SR9)
Traits	Extraversion vs introversion (T1), Openness vs. secretiveness (T2)
Self-image	Self-esteem (SI1)
Motives	Self-presentation (M1), Responsibility (M2)

2.3 Existing Works in Developing Information Privacy Knowledge and Skills

Although researchers agree that Internet users should hold information privacy knowledge, there is scattered findings on what this knowledge should be. Nonetheless, Lavranou and Tsohou (2019) proposed a coherent Information Privacy Common Body of Knowledge which aggregates the knowledge and skills that Internet users should hold to effectively protect their personal data. This framework proposes fives areas of information privacy knowledge and skills:

1. Information Privacy's Stakeholders: Users' skills to identify the roles that are involved in privacy and related to personal information, such as data controller.
2. Infringements against information privacy: the knowledge of types of privacy violations and key facts about them, such as tracking without consent, etc.

3. Strategies to enhance individual's information privacy: the knowledge and skills for applying protective strategies, such as installing protective software etc.
4. Information privacy and organizations: the knowledge of practices used to protect users' privacy, such as privacy seals.
5. Legislative framework: the knowledge of regulatory frameworks on information privacy and the corresponding knowledge of individual rights, pertinent authorities etc.

2.4 Existing Works in Developing Information Privacy Social Role, Traits, Self-image and Motives

In addition to privacy knowledge and skills, researchers formulate a number of other competences that an individual must possess in order to be able to maintain her privacy. These competences belong to four of the six categories that are depicted in Table 1..

The social role refers to the attitudes of the individual and corresponds to eight identified competences. Privacy concerns, regarding how the organizations collect and use personal data, are a component of understanding the concept of privacy (Wagner et al. 2018). Forming a digital identity while being aware of what is being made public and with knowledge and control over the disclosed information is another property of the competent user for protecting her privacy online. Skalkos et al. (2020) research showed that freedom and anonymity while navigating the Internet, self-control over personal information and personal life as well as fear-free browsing on the internet are important values for privacy and Tsohou (2021) classified them into the category of the user's social role. Other researchers have focused on personality traits such as extraversion and openness (Abramova et al. 2017) and self-esteem (Chen et al. 2015) regarding how they influence users' online behavior and especially personal information disclosure and privacy concerns. Finally, researchers have studied how to enable user's motives associated with privacy protection, such as the way one presents herself (Chen et al. 2015) and the responsibility that someone feels to protect others' privacy.

Although the above works present the desired information privacy awareness and competences for Internet users, there is no guidance on how to instill them and how to formulate respective educational interventions. This is a research gap which we aim to address by proposing an educational intervention that may support privacy awareness raising and the development of privacy competent Internet users.

3   Constructivism in the Design of Educational Interventions and Information Privacy Learning

The design of any kind of learning intervention should depend on principles contained in corresponding educational theories, which have been formulated to describe the way in which the individual acquires, processes, understands and retains knowledge. Education sciences rely heavily on three fundamental categories of learning theories. These broad theories have been used mostly to shape the instructional environment: Behaviorism, Cognitivism and Constructivism. These three learning theories are the basis for formulating other, more specialized ones, thereby expanding the scope of each of them.

Constructivism is a theory which asserts that learning is an activity that is individual to the learner (Bada and Olusegun 2015). This learning theory is found in psychology and explains how humans construct knowledge and learn from their experiences (Bereiter 1994). When individuals encounter something new, they need to compare it with their previous ideas, knowledge, and experiences, reframe it, perhaps change their view of what they know, or ultimately reject new information as useless. In order to do that, they need to ask questions, explore new things and recall what they already know. According to the above, constructivists consider the learner as an active procreator in the learning process (Glasersfeld 2013). This view of learning is in direct contradiction with the other learning theories (i.e., Behaviorism) according to which learning is simply the transfer of knowledge from one person to another and not its construction. Modern learning theories are largely new variations of constructivism and researchers tend to agree that learning interventions and teaching environments that meet the needs of constructivism are more effective from those that follow the example of the "traditional" learning theories, such as behaviorism (Bada and Olusegun 2015).

As discussed in Subsect. 2.1, existing educational interventions for information privacy are not based on theoretical grounds related to learning outcomes. This theoretical basis will ensure that sufficient information is provided about the relationships and interconnections between strategies and context in order to achieve better integration. Additionally, the theoretical support allows reliable predictions for the effectiveness of the selected instructional methods (Ertmer and Newby 2013). A key category of constructive educational activities are games of all kinds (Marone 2016). Our literature analysis though shows that such attempts for information privacy are minimal in number and are not based on constructivism. Specifically, *The Watchers* by Raynes-Goldie and Allen (2014) is a hybrid computer/board game, which is played like a conventional tabletop game, augmented by the usage of a computer or tablet. Even though the authors claim several experiential features, its design is not clearly based on a learning theory. *Immaculacy* by Suknot et al. (2014) is an interactive narrative-based game, in which the player is placed in the role of a girl and then the evolvement of the story depends on the decisions of the player regarding the protection of her privacy. *Privacy* by Barnard-Wills and Ashenden (2015) is a card game which aims to enhance the player's privacy literacy by asking them to decide what information they should keep, play or trade with the other players, depending on the character they hold and the storyline. *Friend Inspector* by Cetto et al. (2014) is a memory-like game aiming to raise users' privacy awareness in SNSs, in which players are asked to guess the visibility of a presented disclosed data element within a limited amount of time. Williams et al. (2019) created a smartwatch game to encourage privacy-protective behavior, when using wearable devices, targeting at privacy awareness enhancement, privacy knowledge and behavior change. *Social4School* is game in which children interact in a simulation of a SNS and the main goal is to enhance their perception on privacy issues and to make them aware about the protection of their own and others' personal data (Bioglio et al. 2019). *Leech* is an adventure computer game, that aims to enhance players' understanding and knowledge on privacy policies, through quest-solving gameplay (Pape 2022).

The above efforts highlighted how engaging internet users in games-based learning activities can lead to successful and effective approaches in the development of privacy literacy. However, even though the above games had generic positive results on privacy education, the learning objectives were not clearly specified nor were based on a theoretical framework, so as to promote privacy competent users. We aim to propose alternative educational interventions' design for privacy by adopting a constructive approach, and to clarify two issues. First, constructivism is not a theory of teaching, thus there are no clearly defined tools to guide the design and ensure the effectiveness of teaching (Richardson 2003). Even though research shows that constructive learning is successful in all contexts (Fenstermacher and Richardson 2005), there is lack of a constructivist teaching theory. Second, as an aftermath of the lack of a corresponding pedagogy, the design of constructive learning environments focuses on what constructivism ultimately defines: the way the individual learns. Papaioannou et al. (2022) proposed a constructive pedagogy based on commonly formulated principles of the constructivism learning theory and developed seven principles for privacy learning. Based on these principles, seven requirements for the design of constructive educational interventions were formulated for information privacy: (1) experiential learning, (2) evaluation of alternative solutions and perspectives, (3) authentic tasks, (4) student-centered learning, (5) collaboration, (6) multiple modes of representation and (7) metacognition.

4   Development of the Proposed Framework for Raising Information Privacy Competences

4.1   Games in the Role of Educational Tools

According to Jane McGonigal (2011) people in modern times express a prolonged discontent with everything that happens in real life, while they enjoy satisfaction and positive emotions in a playful environment. Similar views to the above have been used in the literature to propose extending the use of games to other areas of everyday life, such as education, work, culture and sport. Although games' intrusion into peoples' everyday life have been heavily criticized in the past for children's addiction and deprivation of their leisure time, in recent years efforts have been made to highlight the potential they can offer in learning, preserving cultural heritage, acquiring desirable behaviors and cultivating a positive work climate (de Freitas and Liarokapis 2011). Conclusively, it is widely accepted that people are better engaged and motivated for learning through games (Leaning 2015).

4.2   The Effectiveness of Playful Environments for Learning

Given the above and the pervasiveness of electronic games in the field of education, researchers have focused on understanding the causes behind their success. A typical example is the development, diffusion and success of educational games, serious games and virtual worlds games for educational purposes (De Freitas et al. 2010). Serious games are usually defined as computer games that have an educational and learning aspect and are not used just for entertainment purposes (De Freitas et al. 2010). They are usually utilized within a wide range of different conceptual frameworks and objectives. The relevant research field is fragmented, due to its interdisciplinary nature. Therefore, the literature is also scattered in a number of different disciplines, such as pedagogics, psychology and informatics (De Freitas et al. 2010).

Given that teaching and learning through games is considered more effective, even when for difficult topics, relevant research evidence shows that learning through gaming and simulations can help learners that need skills enhancement, support diverse education and bridge the gap between formal and informal learning promoting knowledge acquisition (de Freitas and Oliver 2006).

In the early years of the spread of serious games, researchers were concerned as to whether or not these games were effective compared to other methods and approaches, as there was a serious lack of relevant research in the literature. Several studies in that period have shown positive results in terms of the effectiveness of serious games in teaching/learning and in motivating players/learners from all ages (Kato et al. 2008; Rebolledo-Mendez et al. 2009). Considering that serious games are now implemented through technology, knowledge construction benefit from concepts like immersion, interactivity, narrativity, motivation and engagement. These concepts play an important role in the development of serious games, as technology provide enormous potential in this area (de Freitas and Oliver 2006).

4.3 A New Type of Serious Game: The Escape Room

A promising type of game for educational purposes that has emerged in recent years is the escape rooms. An escape room is a game in which a person or small group tries to "escape" a room in a predetermined time. The room is full of open or hidden challenges that players must solve. When challenges are hidden, players may have to complete puzzles or quizzes in order to proceed. Since their dynamic beginning, these interactive games have evolved to a very high degree of complexity, so as to provide inspiring and enjoyable experiences (Clarke et al. 2017).

A main reason for their popularity is the cooperative nature of the game and its challenges, giving the chance to the players to communicate directly with each other in the real world. Mostly they are narrative-based, and their themes range from mystery, horror, crime and robbery to adventure. The themes create a unique atmosphere that, combined with the time that counts, the kind of challenges, the decoration, and even the background music, emotionally engages the player and arouses his curiosity and excitement (Clarke et al. 2017; Nicholson 2018; Wiemker et al. 2015).

It is worth noting that an escape room mobilizes many skills. There are not many games or activities that involve so many human skills. Wiemker et al. (2015) list this set of skills as follows: Searching, Observation and Discernment, Correlation, Memorization, Math, Words, Pattern Recognition and Compartmentalization. Not all of them are exercised in every escape room, but the number of puzzles, challenges, quizzes and storytelling parameters mobilize a different combination of the skills each time.

4.4 Escape Rooms for Learning

Escape rooms may have any theme related to any field or topic. Thus, escape rooms could be utilized for teaching any subject and it is suitable for all ages. In this paper we argue that the philosophy with which escape rooms are structured makes them ideal teaching and learning tools, according to the constructivism learning theory. This is because, they are usually a cooperative game, which requires teamwork to achieve victory. Further, escape rooms are designed based on specific learning goals, and the players actively learn in an environment they share with their teammates (Nicholson 2018). The knowledge, opinions, and ideas that each teammate can use to solve a puzzle extend the thinking and knowledge of others through interpersonal interaction.

Escape rooms are based on solving logical puzzles and completing sequential tasks. Consequently, this is an activity that triggers thought and motivates the player to use her mind, a characteristic that matches the principles of constructivism. Still, the time limit makes the activity more plausible and actively engages players as the sense of urgency engages interest.

The feature that is common in all escape rooms, and distinguishes them from other games, is the storytelling. Players commonly listening to a narrative, which can be enriched with images and videos, and they become part of the story. All challenges (puzzles, quizzes, locks etc.) are means of aiding the story as they take the player one step further at a time. Still, players can face the consequences of their choices. This can add to the way the game is experienced, make it more plausible and offer the player emotion and motivation to make the right decision. So, this is a type of game in which the player holds a primary role and is not a mere participant. By adding learning parameters to the story, escape rooms can succeed as learning tools, as they offer experiences that cannot be compared to the traditional way of teaching (Nicholson 2018).

Another interesting feature of the escape rooms is the debriefing session. At the end of the game players are informed about the correct course of solving puzzles and challenges. In a learning context, this is an excellent opportunity to expose their views and concerns about the knowledge they received and/or what they found difficult, irrelevant etc. The discussion of learning objectives can be relatively predetermined by game designers and educators, so that the story is properly linked to knowledge (Wiemker et al. 2015). This feature is also key to promote the player's metacognitive skills.

It is important to emphasize again that the key to the success of the goals of an escape room, even in terms of learning, is the story itself and the narrative, not the challenges that the player will solve. This specification is necessary as it is often observed that puzzles, quizzes, locks and coded maps dominate the design of an escape room. This approach is not far from the old educational computer games, which retained their original scenario, in which only encyclopedic questions or arithmetic operations appeared (Nicholson 2018). Focusing only on the puzzles, their level of difficulty and their content, loses the comparative advantage of the escape rooms. This advantage lies in the ability to add the knowledge we want to build into the storyline that the player explores. Veldkamp et al. (2020) provide detailed and useful recommendations in this direction.

Recently, escape room platforms have been exploited systematically for education purposes. *Breakout EDU* (Breakout EDU n.d.; Nicholson 2018) is a platform that assists educators to create puzzle-based boxes for a classroom, aiming students' engagement and the development of learning skills of collaboration, communication, creativity and critical thinking. Huang et al. (2020) created a teaching approach with a digital escape room for the science lesson of 4th elementary grade in order to study the possible effects on students' learning performance, motivation, and problem-solving ability. Several efforts have been made concerning health care professions (Aubeux et al. 2020; Eukel et al. 2017), as well as science, technology, engineering and mathematics (STEM) (Peleg et al. 2019; Queiruga-Dios et al. 2020; Vergne et al. 2020; Watermeier and Salzameda 2019). There are, also, examples of educational escape rooms used in other subjects, such as English language (Urbieta and Peñalver 2019), creative design (Ma et al. 2018) strategic communication (Craig et al., 2020) and information literacy (Pun 2017). Borrego et al. (2017) created an escape room to facilitate the teaching and learning process by motivating university students in the classes of computer networks and information security. Clarke et al. (2017) developed a theoretical framework to support the process of creating an escape room for education. Several studies introduced escape rooms for raising users' security awareness (Oroszi 2019; Mello-Stark et al. 2020; Schneider and Zanwar 2020). Although security awareness is relevant to our research, an extensive review of the works concerning security awareness raising is beyond the scope of our work.

We argue that the characteristics of escape rooms match the seven principles for the design of a constructivist information privacy pedagogy, which were developed by Papaioannou et al. (2022). Escape rooms are ideal for experiential learning, as the storytelling, space objects, and puzzles that need to be solved can give players all the information they need to build new knowledge on their own. Alternative solutions and perspectives also play a key role in the game, as collaboration brings each player face to face with their teammate's point of view and way of thinking. Authentic work is essentially an inherent feature of escape rooms as the game environment can easily simulate the reality and be meaningful for the player, both in terms of space and plot. Moreover, the players feel that they participate in the construction of knowledge, as they are not a passive recipient of knowledge, and they learn through experience. Collaboration is also crucial for the success of players in the game, as in most cases their level of communication depends on whether they will succeed within the required time. Escape rooms are also highly expedient for multiple modes of representation, as they can include all kinds of new technology, such as sound, image, and even virtual or augmented reality elements. Finally, the metacognitive features of the game stand out, as players can understand the way in which they built knowledge, through experience, discovery, collaboration and inner conflict with pre-existing knowledge.

To the best of our knowledge, at present there is no implementation of an escape room that concerns information privacy in the literature, which is the main contributions of this article.

5   An Educational Escape Room for Information Privacy

In this section we propose a constructive educational intervention to cultivate privacy competences, by utilizing an escape room. We created an indicative scenario for an educational escape room for information privacy competences, using the Information Privacy and Personal Data Protection Competency Model for Citizens (Tsohou 2021) as educational goals. We have selected indicative educational goals based on the Competency Model, so that the script has flow, is plausible and motivates the player to get involved in the story. In Table 2., individual points of the script plot are matched with the target competences. In future research we will develop scenarios to cover all the elements of the Competency Model for citizens as educational goals.

Scenario. The players in the introduction to the game receive information about the story, seeing a vision (video) on a screen. In this vision, the main character of the story is introduced, along with scattered images about him and the plot, which end in a tragic finale for him, as his despair leads him to suicide. At the end of the video, one of the players is supposed to wake up as the character in the vision, while the other player(s), who are watching/playing on a different monitor, assume the roles of guardian angels. A clock appears, which starts counting down and invites the players to save the hero, displaying an automated message: "you have one hour to save him".

In the room there are scattered tests of privacy skills, and the players are required to complete them all within one hour. The guardian angel(s) watch the game from the perspective of the main character, who has control of movement and actions, but can see clues around the room that the main character cannot, and they must guide him to accomplish the tasks that will protect his privacy and ultimately save him. Each time they complete one task as a team, it is saved and removed from the "checklist". In case of a wrong choice in the test, the software continues to count down the time and returns the player to the point of the last correct choice, so that she can relive the events and make the right choice. The educational goal is to realize which of his choices were wrong and to come back to correct them. In case the time runs out and the players fails to complete all missions, then a message appears informing that the hero did not survive, as he was led to an irreversible decision for his life. During the game, small time rewards could be given if the player chooses good privacy practices (e.g., to read a privacy policy), although these choices will not be necessary to escape.

The hero of the story reaches the point of suicide because he has successively made very serious mistakes regarding the protection of his personal data and thus completely loses his privacy, reaching despair. Players will find out information that reveals the personality and life of the hero (who he is, what his financial and family situation is, what he wants to hide, etc.). Any correct choice in the tasks will reduce the chance that the hero will reach the despair specified in the dream.

The dream, which the players will see in the introductory part, will be structured in five scenes, where each scene will represent a task. In addition to these five scenes, there will be another one, which will mean the end of the story, in case the players fail to complete their tasks (the suicide). If the players have proceeded with the tasks, having succeeded some and reached a wrong choice, they can see the dream again, but without the scenes concerning the tasks they have solved. Each task is depicted in the dream in two parts: one part related to the content of the task and one with the outcome of the task in case of wrong choice of the players (e.g., unauthorized access of a person to photos with personal moments of the character vs divorce).

In Table 2. we present the proposed five scenes of the storyline which were designed to target the cultivation of relevant privacy competences (alignment with Table 1.).

Table 2. Scenes of the proposed indicative scenario matching competences

	Scenes	IPCM elements
1	**The hero loses friends** and people in his social environment. In an online gaming chat network, there is a recorded chat in which the hero discusses personal problems of a close friend, and she discovers him as she is an anonymous member of the player team. She walks away from him and other people in her group follow when they find out	K3, S1, S7, SR1, SR2, SR9, M2
2	**The hero loses his job.** Players see a video posted on a social network showing the hero talking to someone to claim a job at a competing company. The video was uploaded by an acquaintance who competes for the same position. His employer watches the video and fires him as unreliable	K2, K3, S1, S2, S7, SR1, SR2, T1, T2, SI1
3	**The hero loses money.** The scene shows him receiving a mail that he thinks is from his bank, opens it, follows the link that is attached, enters his login details, and thus authorizes an attacker to extract any amount of money they want. All his money is transferred in real time to a bank of a tax haven	K2, S1, S2, S7, SR1, SR5, SR9
4	**The hero loses his prestige**. He has published years ago lengthy political comments against a particular party. The party searches for his profiles on SNSs, discovers the posts and blacklists him for future collaborations. At the end of the scene, we see the party winning the elections and the hero being irreparably exposed, when he sees his name among other things, in defamatory campaigns of the new government, which present him as an insufficient partner to claim public works construction on behalf of the company he represents	K1, K2, K3, S1, S2, S7, SR1, SR2, SR7, T1, T2, SI1, M1
5	**The hero loses his family.** He often visits websites with content unsuitable for minors and unknowingly installs malware that activates the camera of the laptop and records him. Several months later, his wife receives an email with a video attached showing him inside their house with another woman. They end up in a bad divorce	K2, K3, S1, S2, S4, S5, SR1, SR2, SR5, SR7, SR8, SR9

6 Conclusions

This paper attempts to address the research gap of designing and developing educational interventions for preparing privacy competent Internet users. Even though a lot of research has been done in privacy awareness and knowledge enhancement, they do not specify target learning outcomes, such as the cultivation of certain competences. Moreover, their design is not theoretically founded on educational sciences. Consequently, the following research questions arise: *"What form should an educational approach take to increase competences for privacy protection?"* and *"How can educational approaches increase privacy competences?"*. To answer these questions, we highlight the contribution of game-based learning, and we propose an escape room as a constructive educational intervention for information privacy competences, based on the requirements of the constructivist pedagogy. Furthermore, we present an indicative scenario for the escape room, based on the competences of an internet user to protect her privacy.

The paper offers new contributions which may create implications for several potential interested parties. Designers and developers of educational software and serious games, information security and privacy software companies could rely on our proposal to create constructive escape rooms for privacy. Our work could also be used by researchers as a springboard for further investigation, refinement, and application to other fields of computer science, such as cybersecurity, networks, and programming.

Our future work includes the completion of scenarios to cover all identified privacy competences, the implementation of the software for the escape room, and an empirical validation to test its effectiveness on the learning outcomes based on the concrete educational goals that guided its design. Future research could address ethical issues that may arise in game script formulation, such as preventing behaviors rather than addressing their consequences.

Acknowledgements. This project has received funding from the GSRT for the European Union's Horizon 2020 research and innovation program DEFeND, under grant agreement No 787068.

References

Abramova, O., Wagner, A., Krasnova, H., Buxmann, P.: Understanding self-disclosure on social networking sites - a literature review. In: AMCIS 2017 Proceedings (2017)

Acquisti, A., Gross, R.: Imagined communities: awareness, information sharing, and privacy on the Facebook. In: Privacy Enhancing Technologies, pp. 36–58 (2006)

Aktypi, A., Nurse, J.R.C., Goldsmith, M.: Unwinding Ariadne's identity thread: privacy risks with fitness trackers and online social networks. In: Proceedings of the 2017 on Multimedia Privacy and Security, pp. 1–11 (2017). https://doi.org/10.1145/3137616.3137617

Aubeux, D., et al.: Educational gaming for dental students: Design and assessment of a pilot endodontic-themed escape game. Eur. J. Dental Educ. **24**(3) (2020)

Bada, D., Olusegun, S.: Constructivism learning theory: a paradigm for teaching and learning. J. Res. Method Educ. **5,** 66–70 (2015)

Barnard-Wills, D., Ashenden, D.: Playing with privacy: games for education and communication in the politics of online privacy. Polit. Stud. **63**(1), 142–160 (2015)

Barth, S., de Jong, M.D.T., Junger, M., Hartel, P.H., Roppelt, J. C.: Putting the privacy paradox to the test: online privacy and security behaviors among users with technical knowledge, privacy awareness and financial resources. Telemat. Inform. **41** (2019)

Bereiter, C.: Constructivism, socioculturalism, and popper's world. Educ. Res. **23**(7) (1994)

Bioglio, L., Capecchi, S., Peiretti, F., Sayed, D., Torasso, A., Pensa, R.G.: A social network simulation game to raise awareness of privacy among school children. IEEE Trans. Learn. Technol. **12**(4), 456–469 (2019)

Borrego, C., Fernández, C., Blanes, I., Robles, S.: Room escape at class: escape games activities to facilitate the motivation and learning in computer science. J. Technol. Sci. Educ. **7**(2) (2017)

Breakout EDU - Educational Games. (n.d.). https://www.breakoutedu.com/

Cetto, A., et al.: Friend inspector: a serious game to enhance privacy awareness in social networks. arXiv:1402.5878 [Cs]. http://arxiv.org/abs/1402.5878

Chen, J.V., Widjaja, A.E., Yen, D.C.: Need for affiliation, need for popularity, self-esteem, and the moderating effect of Big Five personality traits affecting individuals' self-disclosure on Facebook. Int. J. Hum. Comput. Interact. **31**(11) (2015)

Clarke, S.J., Peel, D.J., Arnab, S., Morini, L., Keegan, H., Wood, O.: EscapED: a framework for creating educational escape rooms and interactive games to for higher/further education. Int. J. Ser. Games **4**(3) (2017)

Craig, C., Ngondo, P.S., Devlin, M., Scharlach, J.: Escaping the routine: unlocking group intervention. Commun. Teach. **34**(1), 14–18 (2020)

de Freitas, S., Liarokapis, F.: Serious Games: A new paradigm for education? In: Ma, M., Oikonomou, A., Jain, L. (eds.) Serious Games and Edutainment Applications. Springer, London (2011). https://doi.org/10.1007/978-1-4471-2161-9_2

de Freitas, S., Oliver, M.: How can exploratory learning with games and simulations within the curriculum be most effectively evaluated? Comput. Educ. **46**(3) (2006)

De Freitas, S., Rebolledo-Mendez, G., Liarokapis, F., Magoulas, G., Poulovassilis, A.: Learning as immersive experiences: Using the four-dimensional framework for designing and evaluating immersive learning experiences in a virtual world: learning as immersive experiences. Br. J. Edu. Technol. **41**(1), 69–85 (2010)

William, E., et al.: An information-flow tracking system for realtime privacy monitoring on smartphones. ACM Trans. Comput. Syst. **32** (2914). https://doi.org/10.1145/2619091

Ertmer, P.A., Newby, T.J.: Behaviorism, cognitivism, constructivism: comparing critical features from an instructional design perspective. Perform. Improv. Q. **26**(2), 43–71 (2013). https://doi.org/10.1002/piq.21143

Eukel, H.N., Frenzel, J.E., Cernusca, D.: Educational gaming for pharmacy students – design and evaluation of a diabetes-themed escape room. Am. J. Pharm. Educ. **81**(7), 6265 (2017). https://doi.org/10.5688/ajpe8176265

Fenstermacher, G., Richardson, V.: On making determinations of quality in teaching. Teach. Coll. Rec. **107**(1), 186–213 (2005)

Gerber, N., et al.: FoxIT: enhancing mobile users' privacy behavior by increasing knowledge and awareness. Personal Ubiquit. Comput. **13**. https://doi.org/10.1145/3167996.3167999

Ghazinour, K., Matwin, S., Sokolova, M.: Yourprivacyprotector, a recommender system for privacy settings in social networks. ArXiv Preprint **1602**, 01937 (2016)

von Glasersfeld, E.: Radical constructivism. Routledge (2013)

Hatamian, M., Serna, J., Rannenberg, K., Igler, B.: FAIR: fuzzy alarming index rule for privacy analysis in smartphone apps. In: Lopez, J., Fischer-Hübner, S., Lambrinoudakis, C. (eds.) TrustBus 2017. LNCS, vol. 10442, pp. 3–18. Springer, Cham (2017). https://doi.org/10.1007/978-3-319-64483-7_1

Holtkamp, P.: Competency requirements of global software development conceptualization, contextualization, and consequences. Jyväskylä Studies in Computing, Finlandia (2015)

Huang, S.-Y., Kuo, Y.-H., Chen, H.-C.: Applying digital escape rooms infused with science teaching in elementary school: learning performance, learning motivation, and problem-solving ability. Think. Skills Creat. **37**, 100681 (2020)

Joinson, A.N., Reips, U.-D., Buchanan, T., Schofield, C.B.P.: Privacy, trust, and self-disclosure online. Hum. Comput. Interact. **25**(1), 1–24 (2010)

Kato, P.M., Cole, S.W., Bradlyn, A.S., Pollock, B.H.: A video game improves behavioral outcomes in adolescents and young adults with cancer: a randomized trial. Pediatrics **122**(2), e305–e317 (2008). https://doi.org/10.1542/peds.2007-3134

Lavranou, R., Tsohou, A.: Developing and validating a common body of knowledge for information privacy. Inf. Comput. Secur. **27**(5), 668–686 (2019)

Leaning, M.: A study of the use of games and gamification to enhance student engagement, experience and achievement on a theory-based course of an undergraduate media degree. J. Media Pract. **16**(2), 155–170 (2015)

Ma, J.-P., Chuang, M.-H., Lin, R.: An innovated design of escape room game box through integrating STEAM education and PBL principle. In: Proceedings of the Cross-Cultural Design. Applications in Cultural Heritage, Creativity and Social Development, pp. 70–79 (2018)

Malandrino, D., di Salerno, U.: Impact of Privacy Awareness on attitudes and behaviors online. In: CASE, vol. 19 (2012)

Malandrino, D., Petta, A., Scarano, V., Serra, L., Spinelli, R., & Krishnamurthy, B.: Privacy awareness about information leakage: who knows what about me? In: Proceedings of the 12th ACM Workshop on Workshop on Privacy in the Electronic Society, pp. 279–284 (2013)

Marone, V.: Playful constructivism: making sense of digital games for learning and creativity through play. Des. Participat. **9**(3), 20 (2016)

McGonigal, J.: Reality Is Broken: Why Games Make Us Better and How They Can Change the World. Penguin (2011)

Mello-Stark, S., VanValkenburg, M.A., Hao, E.: Thinking outside the box: using escape room games to increase interest in cyber security. Innovations. In: Cybersecurity Education, pp. 39–53. Springer, Cham (2020). https://doi.org/10.1007/978-3-030-50244-7

Nicholson, S.: Creating engaging escape rooms for the classroom. Child. Educ. **94**(1), 44–49 (2018). https://doi.org/10.1080/00094056.2018.1420363

Oroszi, E.D.: Security awareness escape room - a possible new method in improving security awareness of users. In: 2019 International Conference on Cyber Situational Awareness, Data Analytics and Assessment (Cyber SA), pp. 1–4 (2019)

Papaioannou, T., Tsohou, A., Karyda, M., Karagiannis, S.: Requirements for an information privacy pedagogy based on the constructivism learning theory. In: Proceedings of the 17th International Conference on Availability, Reliability and Security (ARES 2022), Vienna, Austria. ACM, New York, NY, USA (2022)

Pape, S.: Challenges for designing serious games on security and privacy awareness. In: Friedewald, M., Krenn, S., Schiering, I., Schiffner, S. (eds.) Privacy and Identity 2021. IAICT, vol. 644, pp. 3–16. Springer, Cham (2022). https://doi.org/10.1007/978-3-030-99100-5_1

Paspatis, I., Tsohou, A., Kokolakis, S.: AppAware: a policy visualization model for mobile applications. Inf. Comput. Secur. **28**(1), 116–132 (2020)

Peleg, R., Yayon, M., Katchevich, D., Moria-Shipony, M., Blonder, R.: A lab-based chemical escape room: educational, mobile, and fun! J. Chem. Educ. **96**(5), 955–960 (2019). https://doi.org/10.1021/acs.jchemed.8b00406

Pun, R.: Hacking the research library: Wikipedia, trump, and information literacy in the escape room at Fresno State. Libr. Q. **87**(4), 330–336 (2017)

Queiruga-Dios, A., Santos Sánchez, M.J., Queiruga Dios, M., Gayoso Martínez, V., Hernández Encinas, A.: A virus infected your laptop let's play an escape game. Mathematics **8**(2), 166 (2020). https://doi.org/10.3390/math8020166

Raynes-Goldie, K., Allen, M.: Gaming privacy: a Canadian case study of a children's co-created privacy literacy game. Surveill. Soc. **12**(3), 414–426 (2014)

Rebolledo-Mendez, G., Avramides, K., de Freitas, S., Memarzia, K.: Societal impact of a serious game on raising public awareness: The case of FloodSim. In: Proceedings of the 2009 ACM SIGGRAPH Symposium on Video Games, pp. 15–22 (2009)

Richardson, V.: Constructivist pedagogy. Teach. College Record **105**(9) (2003)

Schneider, B., Zanwar, T.: CySecEscape–Escape room technique to raise cybersecurity awareness in SMEs. In: The Future of Education International Conference (2020)

Sim, I., Liginlal, D., Khansa, L.: Information privacy situation awareness: construct and validation. J. Comput. Inf. Syst. **53**(1), 57–64 (2012)

Skalkos, A., Tsohou, A., Karyda, M., Kokolakis, S.: Identifying the values associated with users' behavior towards anonymity tools through means-end analysis. Comput. Hum. Behavior Rep. **2**, 100034 (2020). https://doi.org/10.1016/j.chbr.2020.100034

Soumelidou, A., Tsohou, A.: Effects of privacy policy visualization on users' information privacy awareness level: the case of Instagram. Inf. Technol. People **33**(2), 502–534 (2019). https://doi.org/10.1108/ITP-08-2017-0241

Soumelidou, A., Tsohou, A.: Towards the creation of a profile of the information privacy aware user through a systematic literature review of information privacy awareness. Telema. Inform. **61**, 101592 (2021). https://doi.org/10.1016/j.tele.2021.101592

Spencer, L.M., Spencer, P.S.M.: Competence at Work Models for Superior Performance. John Wiley & Sons (2008)

Suknot, A., Chavez, T., Rackley, N., Kelley, P.G.: Immaculacy: a game of privacy. In: Proceedings of the First ACM SIGCHI Annual Symposium on Computer-Human Interaction in Play, pp. 383–386 (2014). https://doi.org/10.1145/2658537.2662971

Tadic, B., Rohde, M., Wulf, V.: CyberActivist: tool for raising awareness on privacy and security of social media use for activists. In: Meiselwitz, G. (ed.) SCSM 2018. LNCS, vol. 10913, pp. 498–510. Springer, Cham (2018). https://doi.org/10.1007/978-3-319-91521-0_36

Tsohou, A.: Towards an information privacy and personal data protection competency model for citizens. In: Fischer-Hübner, S., Lambrinoudakis, C., Kotsis, G., Tjoa, A.M., Khalil, I. (eds.) TrustBus 2021. LNCS, vol. 12927, pp. 112–125. Springer, Cham (2021). https://doi.org/10.1007/978-3-030-86586-3_8

Tsohou, A., Kosta, E.: Enabling valid informed consent for location tracking through privacy awareness of users: a process theory. Computer Law Secur. Rev. **33**(4) (2017)

Urbieta, A.S., Peñalver, A.: Escaping from the English classroom. Who will get out first? Aloma:Revista de Psicologia,Ciències de l'educació ide l'esport Blanquerna, **37**(2) (2016)

Veldkamp, A., van de Grint, L., Knippels, M.C.P., van Joolingen, W.R.: Escape education: a systematic review on escape rooms in education. Educ. Res. Rev. **31** (2020)

Vemou, K., Karyda, M., Kokolakis, S.: Directions for raising privacy awareness in SNS platforms. In: Proceedings of the 18th Panhellenic Conference on Informatics, pp. 1–6 (2014)

Vergne, M.J., Smith, J.D., Bowen, R.S.: Escape the (Remote) classroom: an online escape room for remote learning. J. Chem. Educ. **97** (2020)

Wagner, A., Wessels, N., Buxmann, P., Krasnova, H.: Putting a price tag on personal information— a literature review. In: Hawaii International Conference on System Sciences (2018). http://hdl.handle.net/10125/50362

Watermeier, D., Salzameda, B.: Escaping boredom in first semester general chemistry. J. Chem. Educ. **96**(5), 961–964 (2019)

Wiemker, M., Elumir, E., Clare, A.: Escape room games. Game Based Learn. **55** (2015)

Williams, M., Nurse, J.R.C., Creese, S.: (Smart)Watch out! encouraging privacy-protective behavior through interactive games. Int. J. Hum. Comput. Stud. **132**, 121–137 (2019). https://doi.org/10.1016/j.ijhcs.2019.07.012

Winterton, J.: Competence across Europe: highest common factor or lowest common denominator? J. Eur. Ind. Train. **33**(8/9). 681–700 (2019)

Mitigating Sovereign Data Exchange Challenges: A Mapping to Apply Privacy- and Authenticity-Enhancing Technologies

Kaja Schmidt[1]([⊠]) ⓘ, Gonzalo Munilla Garrido[2] ⓘ, Alexander Mühle[1] ⓘ,
and Christoph Meinel[1]

[1] Hasso Plattner Institute, Potsdam, Germany
{kaja.schmidt,alexander.muehle,christoph.meinel}@hpi.de
[2] Technical University of Munich, Munich, Germany
gonzalo.munilla-garrido@tum.de

Abstract. Harmful repercussions from sharing sensitive or personal data can hamper institutions' willingness to engage in data exchange. Thus, institutions consider Authenticity-Enhancing Technologies (AETs) and Privacy-Enhancing Technologies (PETs) to engage in Sovereign Data Exchange (SDE), i.e., sharing data with third parties without compromising their own or their users' data sovereignty. However, these technologies are often technically complex, which impedes their adoption. To support practitioners select PETs and AETs for SDE use cases and highlight SDE challenges researchers and practitioners should address, this study empirically constructs a challenge-oriented technology mapping. First, we compile challenges of SDE by conducting a systematic literature review and expert interviews. Second, we map PETs and AETs to the SDE challenges and identify which technologies can mitigate which challenges. We validate the mapping through investigator triangulation. Although the most critical challenge concerns data usage and access control, we find that the majority of PETs and AETs focus on data processing issues.

Keywords: Sovereign data exchange · Technology mapping · Privacy-enhancing technologies · Authenticity-enhancing technologies

1 Introduction

Companies seeking growth collect and analyze increasing amounts of data to innovate and improve their products and services. Thereby, data becomes a valuable resource [12]. However, despite the upside potential of collecting more data, institutions are reluctant to share (often sensitive) information with third parties as they may lose control over who and how their data is accessed, used, or distributed [4,29]. European initiatives (e.g., Gaia-X [6]) were formed to entice the industry into considering data sharing without risking unauthorized distribution or misuse of data (i.e., data sovereignty). We refer to *Sovereign Data*

S. Katsikas and S. Furnell (Eds.): TrustBus 2022, LNCS 13582, pp. 50–65, 2022.
https://doi.org/10.1007/978-3-031-17926-6_4

Exchange (SDE) as the ability of a digital subject to share their data with third parties without compromising their data sovereignty, i.e., the self-determination over accessing, processing, managing, or securing their data.

As institutions work towards SDE, Privacy-Enhancing Technologies (PETs) and Authenticity-Enhancing Technologies (AETs) receive increasing attention. PETs manage or modify data to protect sensitive personal information [35] and AETs enhance authenticity, and integrity of data and information in a system [52] to incentivize data sharing [67]. Some PETs can also be AETs, but not all AETs are PETs and vice versa [52]. Nonetheless, we treat PETs and AETs jointly. However, researchers and practitioners struggle to understand the technologies due to their technical complexity, low maturity, the wide range of possible variations and combinations, and potential economic risks [52,80,81]. While researchers focused on SDE challenges [4,44,81] or capabilities of PETs and AETs [20,62], we are the first to help practitioners select PETs or AETs according to the challenges in SDE use cases.

The paper is structured as follows: First, we identify the challenges of SDE with a Systematic Literature Review (SLR) and Expert Interviews (EIs) (Sect. 2.1). Consecutively, we map PETs and AETs against the challenges they can tackle via investigator triangulation (Sect. 2.2). We provide four contributions. We (1) compile a list of relevant *SDE challenges* (Sect. 4.1), (2) thoroughly research PETs and AETs relevant to mitigate SDE challenges, displayed in a *technology classification* (Sect. 3), (3) map PETs and AETs against the SDE challenges in a *technology mapping* (Sect. 4.2), and (4) outline five *key findings* (Sect. 5). The related work (Sect. 6) and conclusion (Sect. 7) wrap up the paper. Overall, our study synthesizes previous research in an actionable way: the technology mapping is a challenge-oriented approach supporting the identification of appropriate PETs and AETs to tackle SDE challenges.

2   Research Methods

2.1   Methodologic Triangulation

RQ1: Based on researchers and industry stakeholders, what challenges hinder SDE? We conduct an SLR and complement the findings through EIs to answer RQ1. We consolidate the results into 13 challenges of SDE (Table 3), classified into organizational issues, data processing and publishing, and infrastructure challenges (Table 4).

Systematic Literature Review. The SLR process consists of three phases [77, 78], which are documented to increase transparency and reproducibility of the SLR. First, we define a search strategy by choosing search terms and databases. We apply the search term *"data sovereign*""* to multiple databases (Table 1) to ensure a coverage of software engineering and information systems literature. As of April 2021, we identified 205 search hits. Second, we define inclusion criteria and exclusion criteria. At first, we included all publications satisfying formal requirements (e.g., unique English journal and conference hits). Then, we exclude articles based on titles and abstracts (e.g., unrelated data sovereignty

Table 1. Selection process of the SLR

Selection criteria	ACM Digital Library[a]	IEEE Xplore Digital Library[b]	Web of Science[c]	Total
Total search hits	71	49	85	205
Unique journal/conference hits	49	64	26	139
Title and abstract hits	17	14	15	46
Full-text hits	5	4	5	14
Backward search final hits	–	–	–	2
Total final hits	–	–	–	**16**

[a]https://dl.acm.org/, [b]https://ieeexplore.ieee.org/, [c]https://www.webofknowledge.com/

subject, too broad focus) followed by exclusion based on full-test analysis (e.g., no data exchange focus, generic challenges of technologies). We exclude 191 hits, resulting in a subset of 14 final hits (6.8% relevance rate). We identified 2 more hits from the backward search. The final hits were published between 2014 and 2021. Third, we analyze the final hits using content analysis [48]. We extract, summarize, thematically compare, and generalize text passages describing SDE challenges. Overall, 13 challenges are highlighted (Table 3).

Expert Interviews. Following the guidelines by Runeson and Höst [63] and Morse [50], we conduct EIs to enrich the findings of the SLR with opinions from the industry and reduce bias from the SLR. The empirical method consists of two parts. First, we devise a questionnaire[1] with independently reviewed questions. Second, participants are recruited by contacting 54 individuals who have co-authored journal and conference papers on SDE (e.g., [12]), who have worked directly or indirectly on data sovereignty projects, and who come from different institutions. Six individuals,[2] of whom none co-authored this paper, agreed to participate in April and May 2021. Two interviews were conducted as synchronous, semi-structured online interviews [63], while the remaining interviews occurred via E-Mail correspondence. All questions were fully answered. Participants provided consent and were informed about organizational matters (e.g., study purpose, right to withdraw responses). The findings are summarized before concluding the interviews to counter selection bias and avoid misinterpretation [63]. Similar to the data analysis in the SLR, the expert interview responses are paraphrased and analyzed using content analysis. Then, the challenges are consolidated with challenges from the SLR (Table 3).

2.2   Investigator Triangulation

RQ2: *Which PETs and AETs have the potential to mitigate the SDE challenges identified through RQ1?* We apply investigator triangulation [75] to map which PETs and AETs from the technology classification (Sect. 3) can mitigate the SDE challenges from RQ1. The evaluation of the individual technologies was

---

[1] Questionnaire: https://anonymous.4open.science/r/trustbus2022-5C26/.
[2] Three IT project managers, one computer science and two security researchers.

conducted by each researcher to decrease bias in the evaluation stage and contribute to the findings' internal validity [75]. First, every researcher selects one technology from the technology classification (Sect. 3). Second, each researcher individually goes through the challenges from RQ1 one by one, and gauges whether the technology fully or partially addresses the challenge on a theoretical or technical basis. Third, the researchers' evaluations are consolidated with one another through rigorous discussions, and recorded in writing.[3] The final results are summarized in a technology mapping (Table 5).

3   Technology Classification

PETs *manage* data through privacy principled architectures and policies or *modify* data with heuristics or mathematical privacy guarantees to protect personal or sensitive information while minimally disturbing data utility [35,52,76]. In contrast, AETs support and improve the assessment of authenticity and integrity of data in a digital system [52]. AETs thus facilitate the assessment of trust and confidence between parties [18,40] and ensure accountability and compliance [7].

Inspired by [20,52], we classify PETs and AETs into five layers. The layers correspond to data communication, storage, processing, verification, and sovereignty (Table 2). The *data communication layer* and *storage layer* address how to transfer and store data, respectively, in a secure and trustworthy manner. The *data processing layer* includes technologies that enhance the privacy of the data input, its computation, and output [52,76]. The *data verification layer* is concerned with certification of identities, properties of individuals, and attributes of datasets or resources. Lastly, the *sovereignty layer* includes mechanisms that enforce usage control and privacy protection on a policy basis [24].

4   Challenges and Technology Mapping

4.1   SDE Challenges (RQ1)

The concept matrix (Table 3) lists 13 SDE challenges identified from analyzing the final 16 literature hits and interviewing 6 experts. The SDE challenges are described and grouped into organizational issues, data processing and publishing, and infrastructure issues (Table 4). The *organizational challenges* relate to institutions' uncertainties regarding legislation, technology standards, opportunity costs of data exchange, and by extension, SDE. The *data processing and publishing challenges* are primarily concerned with protecting digital subjects' personal interests and privacy in data exchange. Lastly, *Infrastructure challenges* relate to data security and privacy beyond processing and publishing data. Specifically, the challenges deal with implementation issues regarding data access and usage control, strengthening trust in the infrastructure and amongst data exchange participants, as well as enforcing accountability and auditability. Fewer challenges were mentioned by the experts than by the final literature hits.

---

[3] Assessment tables: https://anonymous.4open.science/r/trustbus2022-5C26/.

Table 2. Overview of PETs and AETs

Technology
Data Communication Layer
Encryption. A core cryptographic technique to send data between entities, which only intended recipients can decrypt [23]
Anonymous Routing. As the backbone of The Onion Router (TOR) [25], the protocol anonymously relays messages through a distributed network while being resistant to eavesdropping and traffic analysis [74]
Data Storage Layer
Decentralized Storage. Distributed Hash Tables (DHTs) are at the core of decentralized storage systems and can be used to store and retrieve data distributed across the nodes of a Peer-to-Peer (P2P) network [58]
Searchable Encryption (SE). SE supports search functionality on the server-side without decrypting data and losing data confidentiality [10,71] through Searchable Symmetric Encryption (SSE) or Public-Key Encryption with Keyword Search (PEKS)
Data Processing Layer
Homomorphic Encryption (HE). Arithmetic operations directly on ciphertext, such that only authorized entities can decrypt and access the output of the operations [16,47]
Secure Multi-Party Computation (MPC). Data exchange participants can jointly compute functions on data without revealing their data inputs to other participants. Popular implementations are based on secret-sharing [69] and garbled circuits [79].
Federated Learning (FL). Multiple participants can train machine learning models collaboratively over remote devices [42,46], i.e., participants keep their data localized and only share local model updates with a coordinating central server. Thus, data privacy is enhanced as data never leaves the data owner's device [46,60]
Trusted Execution Environment (TEE). Secure memory areas physically isolated from the device's operating system and applications [37,55]. Unique encryption keys are associated with hardware, making software tampering as challenging as hardware tampering [52]
Differential Privacy (DP). Algorithms fulfilling this privacy definition enhance privacy by adding randomized noise to an analysis, such that its results are practically identical with or without the presence of an individual data subject [26], providing plausible deniability
k-Anonymity. This privacy model uses syntactic building blocks (suppression and generalization) to transform a dataset such that an individual cannot be distinguished from at least $k - 1$ others in the dataset [64,65,73]
Pseudonymization. Replaces identifiers with pseudonyms via encryption, hash functions, or tokenization to decreases the linkability to individuals [54,70].
Verification Layer
Distributed Ledger Technology (DLT). Distributed and tamper-proof database, where the state is stored on multiple nodes of a cryptographically secured P2P network [11]. The state is updated on all nodes using a consensus algorithm
Verifiable Credential (VC). VCs are sets of verifiable claims that can prove the authenticity of attributes or identities [49]. The standardized digital credentials use Decentralized Identifiers (DIDs) and Digital Signatures (DSs) to form attestation systems
Zero-Knowledge Proof (ZKP). Cryptographic protocol allowing to authenticate knowledge without revealing the knowledge itself [31,32]. ZKPs and can provide data authenticity, identity authenticity, and computational integrity.
Sovereignty Layer
Privacy-by-Design (PbD). The seven-step guidelines can protect privacy in systems' designs by acknowledging privacy within risk management and design processes (e.g., privacy by default settings, end-to-end security of personal data) [13,34]
Privacy Policy (PP). PPs (e.g., through smart contracts [52] or on a contractual basis [12]) embody privacy requirements and guidelines of data governance models [72], such that different policies can be applied to different data consumers

Table 3. Concept matrix of identified SDE challenges

Data Source	Managing Jurisdictions (C1)	Missing Standards (C2)	Reluctance (C3)	Ensuring Data Privacy (C4)	Ensuring Data Quality (C5)	Ensuring Computational Privacy (C6)	Interoperability (C7)	Minimizing Computational Complexity (C8)	Inter-Organizational Trust (C9)	Cyber Security & Trust in Infrastructure (C10)	Data Provenance (C11)	Data Usage & Access Control (C12)	Auditability (C13)
Systematic Literature Review													
Panhuis et al. [59]	✓	✓	✓	✓	✓		✓		✓			✓	
Lablans et al. [43]			✓			✓	✓						
Ahmadian et al. [2]	✓			✓				✓		✓	✓		✓
Bennett et al. [5]				✓	✓					✓		✓	
Brost et al. [9]									✓	✓		✓	
Demchenko et al. [21]		✓											
Celik et al. [14]	✓		✓			✓							
Cuno et al. [19]	✓	✓	✓										
Otto et al. [57]				✓	✓		✓	✓	✓	✓	✓	✓	
Sarabia-Jacome et al. [66]							✓						
Zrenner et al. [81]				✓			✓	✓				✓	
Gil et al. [29]									✓		✓	✓	
Lee et al. [45]				✓									
Nast et al. [53]						✓						✓	
Andreas et al. [3]	✓			✓			✓	✓					
Grünewald et al. [33]		✓		✓						✓	✓		
Count from SLR	5	4	4	8	3	3	6	4	4	5	4	7	1
Expert Interviews													
Interviewee 1 [I1]				✓			✓				✓	✓	
Interviewee 2 [I2]									✓			✓	
Interviewee 3 [I3]						✓						✓	
Interviewee 4 [I4]		✓	✓		✓	✓	✓					✓	✓
Interviewee 5 [I5]				✓		✓			✓		✓	✓	✓
Interviewee 6 [I6]		✓									✓	✓	
Count from Interviews	-	2	1	2	1	3	2	-	2	-	3	6	2
Total No. References	**5**	**6**	**5**	**10**	**4**	**6**	**8**	**4**	**6**	**5**	**7**	**13**	**3**

The individual challenges are derived through content analysis following Sect. 2.1. For example, *missing standards* (C2) is mentioned by three literature sources and two interviewees. Researchers highlighted the challenge of missing standardizing guidelines on data sharing [59], standards regarding the publishing of smart city data [19], or end-to-end standards [I4]. More generally, [21,33] and [I6] encounter challenges of missing standardization in inter-system communication, transparency information items, and reusable open-source solutions to technically ensure compliance with the General Data Protection Regulation (GDPR). The limitations caused by missing data format and architecture challenges were summarized into one challenge. As standardization processes typically require collaborative working groups (e.g., ISO), C2 was considered an organizational challenge. The remaining challenges were derived similarly.

4.2   Technology Mapping (RQ2)

As described in Sect. 2.2, PETs and AETs (Sect. 3) are evaluated based on their potential to mitigate the SDE challenges identified in RQ1. We summarize the

Table 4. Identified SDE challenges (RQ1)

No.	SDE challenge description
Organizational challenges	
C1	**Managing Jurisdictions.** Uncertainties about interpreting and implementing legal guidelines in technological infrastructures [2] (e.g., data sharing, copyright, data ownership, or personal data) in different jurisdictions [3,14,59]. Varying regulations for (at times hard to distinguish) personal and anonymized datasets cause further complications [19,59]
C2	**Missing Standards.** The definition and implementation of suitable technological solutions for SDE use cases lack standards [I2], e.g., lack of end-to-end standards [I5] for data exchange, use, and replication [19], technologically translating data protection regulations [21,33], data formats [59], and proving data integrity and authenticity [I6]
C3	**Reluctance.** Despite trustworthy technical guarantees, organizations hesitate to engage in data exchange because the terminology "sharing/exchange" suggests that raw data is transferred to third parties [I4]. Organizations lack incentives [59], as the benefits of data exchange rarely outweigh high opportunity costs (e.g., privacy and (sensitive) data breaches) [43]
Data processing and publishing challenges	
C4	**Ensuring Data Privacy.** Data privacy requires data to be manipulated and protected (e.g., minimize data disclosure, secure data storage) [5,57] to prevent unauthorized third party to draw conclusions about individual entities [I4] and following privacy regulations (e.g., GDPR) [33,45]
C5	**Ensuring Data Quality.** Privacy-sensitive data should be usable after anonymization. Thus, the challenge is to enable C4 while preserving data usability [57], i.e., to meet data quality requirements of data consumers [I1], to incentivize data exchange
C6	**Ensuring Computational Privacy.** Computational privacy refers to keeping the metadata, and semantic views of a data transaction secret [43]: who transferred data to whom, what algorithm was applied, and which dataset was accessed [I4]. Thus, ensuring computational privacy is a challenge that extends C4 beyond the dataset's content
C7	**Interoperability.** Technological interoperability standardizes interactions between parties, e.g., authentication, authorization, or data exchange agreement protocols [57]. Semantic interoperability describes datasets through standardized metadata schemes stored in central metadata repositories [59] to facilitate the search of heterogeneous or non-standardized datasets or handle datasets that have been normalized differently [I1]
C8	**Minimizing Computational Complexity.** Data processing and publishing needs scalable and affordable SDE implementations [81]. Data exchange requires low latency, computational complexity, and parallel processing to limit calculation time and memory usage [3] to make privacy-preserving alternatives affordable and usable [57]
Infrastructure challenges	
C9	**Inter-Organizational Trust.** Parties do not share data unless they trust the other parties [29] (e.g., platform operators). Companies are more likely to trust others if they have had contact with one another before [I2], apply unenforceable, and untrackable soft agreements [I4], or operate in a data ecosystem with a trusted root of trust [I1]
C10	**Cyber Security & Trust in Infrastructure.** SDE must ensure *confidentiality*, *integrity*, *availability*, and *resilience* of data and the infrastructure (soft- and hardware). Specifically, the infrastructure must be transparent [2,33], data communication and storage secured, and unauthorized access prohibited [9]
C11	**Data Provenance.** For data exchange, data provenance is often enabled through blockchain-based technologies, logging, and monitoring transactions [57] to establish *accountability* [2], integrity, non-manipulation of data [I1], and authentication processes [29,33]. Decisions include how long data is stored, what data is stored, and how privacy protection is handled [33]
C12	**Data Usage & Access Control.** Legal and technological data control is lost after data is exchanged [19,81], i.e., there is a high risk of unauthorized copying, redistribution, or reusing data for unintended purposes [I5] by unauthorized parties [I6]. Liabilities and accountability must be transferred with data to revoke access rights, specify a data storage location, intervene manually, or specify purpose- and time-based access rights to data [9]
C13	**Auditability.** Proving the legitimacy of claims about complying with data-related guidelines [2] (e.g., proving that datasets have been anonymized by applying PETs, proving that consent and policy agreements are followed) is required for accountability

evaluation in a *technology mapping* (Table 5). For example, DP achieves anonymization in a dataset by adding randomized noise to the data. We find that DP does not support any organizational challenge as it only addresses datasets' contents. Thus, challenges C1, C2, and C3 are not addressed, and the cells remain blank. Contrarily, DP addresses data processing and publishing challenges. Specifically, the technique ensures data privacy (C4) while maintaining data quality (C5) and providing parameter that helps balancing privacy and accuracy. However, data quality is reduced as data is perturbed, meaning that C5 is only partially maintained. Furthermore, DP does not have high computational complexity compared to other processing layer technologies and thus mitigates C8. DP can affect inter-organizational trust (C9) in the peripheral and provide the parameters used for DP to partially fulfill compliance verification (C13).

5 Discussion

The following section draws key findings (KF) and implications from the SDE challenges (RQ1) and the technology mapping (RQ2), outlines the study's limitations, and proposes research questions for future work.

KF1: The final literature hits can be categorized into three thematic research streams: security and challenges of data exchange, International Data Spaces (IDSs), and design of sovereignty layer technologies. The most recent research stream focuses on sovereignty layer technologies, suggesting an upward trend in the research area of SDE. The first research stream more generally focuses on the security and/or challenges of data exchange (6 studies, published 2014–2021) and the IDS stream describes implementations and use cases for the Data Spaces Association (IDSA) data governance framework (5 studies, published 2018–2020), e.g., urban [19] or seaport data spaces [66]. The third research stream focuses on technologies of the sovereignty layer (e.g., privacy languages, privacy policies, consent frameworks) and contains 5 publications since 2019.

KF2: No single PET or AET addresses all SDE challenges, suggesting that PETs or AETs must be carefully chosen, combined, further developed, or complemented with new techniques. Practitioners must carefully select their primary challenge to choose suitable technologies for SDE use cases. For example, data provenance (C11) requires logging and surveillance to establish accountability. While logging can support C11, auditability (C13), and inter-organizational trust (C9) [57], it impedes ensuring data privacy (C4) and computational privacy (C6) and could increase the reluctance (C3) of data exchange participants.

KF3: A large portion of PETs and AETs address data processing and publishing challenges (C4–C8), and barely address organizational challenges (C1–C3). This suggests that the research field of SDE is still maturing and has not yet established best practices. While data processing and publishing challenges are the core technical concepts of SDE, organizational challenges depict issues that arise in productive SDE systems. However, as best practices for data processing

Table 5. Mapping PETs and TETs to SDE challenges (RQ2)

Technology	Managing Jurisdictions (C1)	Missing Standards (C2)	Reluctance (C3)	Ensuring Data Privacy (C4)	Ensuring Data Quality (C5)	Ensuring Computational Privacy (C6)	Interoperability (C7)	Minimizing Computational Complexity (C8)	Inter-Organizational Trust (C9)	Cyber Security & Trust in Infrastructure (C10)	Data Provenance (C11)	Data Usage & Access Control (C12)	Auditability (C13)
Data Communication Layer													
Encryption										✓	·	·	
Anonymous Routing						·				·			
Data Storage Layer													
Decentralized Storage					✓				·		·	✓	
Searchable Encryption				·	✓	✓	·		·	✓			✓
Data Processing Layer													
Homomorphic Encryption				✓	✓	·			·	✓		·	
Secure Multiparty Computation				✓	✓	·			·			✓	·
Federated Learning				✓	✓	·		·	·	·		✓	
Trusted Execution Environment				✓	✓	✓		·		✓	·		
Differential Privacy				✓	·			✓	·				·
k-Anonymity				✓	·			·	·				·
Pseudonymization				·	✓			✓					
Verification Layer													
Distributed Ledger						·		·	✓	·	✓		
Verifiable Credential	✓			·		·	·	✓	✓	·	✓	·	
Zero-Knowledge Proof				✓	✓	·		·	✓	·	✓		✓
Sovereignty Layer													
Privacy-by-Design		·	·	*	*	*		*	*	*	*	*	*
Privacy Policies	·		·	*	*	*	*	*	*	*	*	✓	*
Total Count (excl. *)	**1**	**2**	**2**	**10**	**11**	**8**	**2**	**8**	**10**	**11**	**5**	**7**	**5**

✓ : addressed technically and theoretically
· : addressed technically or theoretically (not both), or only under specific circumstances
* : characteristics depend on the selected technologies to fulfill the requirements

and publishing challenges are still in development [I2,I5], there are no productive SDE systems, rendering the organizational challenges redundant and peripheral. We deduce that research on organizational challenges will likely gain importance once data processing and publishing challenges have been thoroughly explored and state-of-the-art solutions are presented. The first indications are the growing interest in policy and enforcement research [19] (e.g., Gaia-X [6], IDSA [56]) to leverage data sharing while protecting sensitive data [I4].

KF4: Data usage and access control (C12) is the most critical challenge for researchers and experts. Mechanisms must inhibit the unauthorized redistribution of data, revoke access rights, and transfer liabilities with data [I5]. The fundamental technical problem is to introduce data sharing without allowing third parties to use privacy-sensitive or confidential data for non-designated purposes [I4]. Access rights remain a major technological challenge given the fluid nature of data [19,81]. Thus, more attention should be given to implement C12 (e.g., FL can enable C12 because data is never shared with a third party, MPC can support C12 by exchanging encryption keys regularly).

KF5: Given that the research field of SDE is still maturing (KF3), challenges that are currently rarely mentioned by researchers and practitioners, i.e., auditability (C13), managing jurisdictions (C1), and minimizing computational complexity (C8), are likely to gain importance as PETs and AETs face real world barriers. Although the relevance of C1, C8, and C13 is recognized [2,59], the challenges remain on the peripheral of PETs and AETs research. Languages to interpret jurisdictions must be refined [28,33] and the computational complexity of technologies must be managed to become established technologies [3,52]. We thus anticipate an increased need for research on C1, C8, and C13.

Limitations. Despite the rigorous research design and process, several limitations to SLRs and interviews may undermine the effectiveness, completeness, subjectivity, or accuracy of the findings. To minimize bias and strengthen the findings, we defined inclusion and exclusion criteria ex-ante and applied triangulation for the SLR. While the sample of interviewees was low, we employed the EIs as supporting evidence for the SLR and not as standalone findings. We tried balancing the varying levels of expertise of interviewees by sharing the questionnaire prior to the interview for preparational purposes. Lastly, we note that, while we applied investigator triangulation to map technologies with SDE challenges, we have not tested the usability of the mapping.

Future Work. We encourage researchers and practitioners to use the concept matrix (Table 3) and technology mapping (Table 5) as a starting point to locate SDE challenges, identify research gaps, or choose suitable PETs or AETs for SDE use cases. For example, interoperability (C7) is only addressed by SE, VC and PP However, *what solution propositions exist to mitigate interoperability challenges of data and system architectures on a large scale? Similarly, how can the auditability challenge be mitigated without compromising data privacy and computational privacy goals?* Additionally, researchers and PETs or AETs experts can refine the SDE challenges and the technology mapping to clarify the body of knowledge: *Which PETs and AETs have the potential to mitigate or worsen SDE challenges? What are best practices to implement PETs and AETs in SDE use cases? Which PETs and/or AETs act as complements or supplements in the context of SDE?* Finally, practitioners can use the technology mapping as a starting point to implement SDE use cases, and thereby help evaluate the usability of the technology mapping in a real-world context.

6 Related Work

Several surveys presented an overview of PETs and AETs [1,22,68] or taxonomies with the technologies' qualities (e.g., cryptographic foundation, data handling requirements) [30,36], adoption challenges (e.g., social, technical) [41, 80], or PETs in different contexts (e.g., blockchain) [39,61]. However, while these

60   K. Schmidt et al.

surveys presented extensive research to understand and contextualize PETs, they did not focus on data exchange. Only few studies investigated PETs and AETs in the context of data movement [52,62]. Although the studies presented useful tables mapping PETs and AETs against characteristics (e.g., confidentiality), the privacy-oriented mappings did not focus on supporting the application of PETs and AETs. Similarly, a data sovereignty challenge-oriented mapping [44] superficially presented solution approaches instead of concrete PETs and AETs.

Others focused on the application of PETs and AETs. There are handbooks describing how to design privacy-preserving software agents [8], legally implement privacy requirements [20], or how to adopt PETs using a question-based flowchart [15]. Similarly, there are overviews of business use cases for which PETs can be applied [17,27,38]. Others [51] outlined which PETs and AETs can address challenges of value chain use cases. However, the use case-oriented mappings did not support the implementation of PETs and AETs. More concretely, Papadopoulos et al. [60] presented a use case that implements FL to meet the privacy and trust requirements of the involved participants. They demonstrated that the challenges of SDE are addressable, but did not provide a framework to support the implementation of similar endeavors. Furthermore, there exist data ecosystem reference architectures (e.g., Gaia-X [6], IDSA [56]) and policy frameworks [81]. Although the reference architectures have presented holistic solution propositions to SDE challenges, the data ecosystems are not yet practicable.

Overall, studies either investigated the potentials of PETs and AETs without a specific focus on SDE, or addressed SDE challenges with an individual PET or AET. No study proposes a technology mapping to help practitioners and researchers identify suitable PETs and AETs when implement SDE use cases. However, this work is necessary to support researchers and practitioners in understanding and integrating the technologies in practice.

7   Conclusion

We structure the landscape of SDE challenges and identify suitable mitigating technologies, thereby guiding the implementation of SDE use cases and informing researchers of potential future research areas. A two-pronged approach was pursued. First, we identified 13 SDE challenges through a SLR and EIs (Table 4). Second, we proposed PETs and AETs to mitigate the identified SDE challenges using investigator triangulation, summarized in a technology mapping (Table 5). The technology mapping synthesizes previous research in an actionable way for practitioners and researchers by presenting a challenge-oriented approach that supports the identification of appropriate PETs and AETs to tackle SDE challenges – regardless of the use case. No single technology mitigates all SDE challenges, indicating that PETs and AETs can be combined, further investigated by researchers, or complemented with new solutions. In particular, we suggest focusing on access control and jointly facilitating auditability and data privacy.

References

1. Adams, C.: Introduction to Privacy Enhancing Technologies: A Classification-Based Approach to Understanding PETs. Springer, Cham (2021). https://doi.org/10.1007/978-3-030-81043-6
2. Ahmadian, A.S., Jürjens, J., Strüber, D.: Extending model-based privacy analysis for the industrial data space by exploiting privacy level agreements. In: Proceedings of the 33rd Annual ACM Symposium on Applied Computing, pp. 1142–1149 (2018)
3. Andreas, A., et al.: Towards an optimized security approach to IoT devices with confidential healthcare data exchange. Multimed. Tools Appl. **80**(20), 31435–31449 (2021). https://doi.org/10.1007/s11042-021-10827-x
4. Bastiaansen, H.J., Kollenstart, M., Dalmolen, S., van Engers, T.M.: User-centric network-model for data control with interoperable legal data sharing artefacts: improved data sovereignty, trust and security for enhanced adoption in interorganizational and supply chain in applications. In: 24th Pacific Asia Conference on Information Systems, Dubai, UAE, pp. 1–14. AIS (2020)
5. Bennett, C., Oduro-Marfo, S.: GLOBAL privacy protection: adequate laws, accountable organizations and/or data localization? In: 2018 ACM International Joint Conference on Pervasive and Ubiquitous Computing, pp. 880–890 (2018)
6. BMWi: Project GAIA-X: a federated data infrastructure as the cradle of a vibrant European ecosystem. Technical report, Federal Ministry for Economic Affairs and Energy (BMWi), Berlin, Germany (2020)
7. Bordel, B., Alcarria, R.: Trust-enhancing technologies: blockchain mathematics in the context of Industry 4.0. In: Advances in Mathematics for Industry 4.0, Amsterdam, Netherlands, pp. 1–22. Academic Press (2021)
8. Borking, J.J., Raab, C.D.: Laws, PETs and other technologies for privacy protection. J. Inf. Law Technol. **1**, 1–14 (2001)
9. Brost, G., Huber, M., Weiß, M., Protsenko, M., Schütte, J., Wessel, S.: An ecosystem and IoT device architecture for building trust in the industrial data space. In: Proceedings of the 4th ACM Workshop on Cyber-Physical System Security, Incheon, Republic of Korea, pp. 39–50. ACM (2018)
10. Bösch, C., Hartel, P., Jonker, W., Peter, A.: A survey of provably secure searchable encryption. ACM Comput. Surv. **47**(2), 1–51 (2014)
11. Butijn, B.J., Tamburri, D.A., van den Heuvel, W.J.: Blockchains: a systematic multivocal literature review. ACM Comput. Surv. **53**(3), 1–37 (2020)
12. Cappiello, C., Gal, A., Jarke, M., Rehof, J.: Data ecosystems: sovereign data exchange among organizations. Dagstuhl Rep. **9**(9), 66–134 (2020)
13. Cavoukian, A.: Privacy by design: the 7 foundational principles. Technical report, Information and privacy commissioner of Ontario, Canada (2009)
14. Celik, Z.B., Acar, A., Aksu, H., Sheatsley, R., McDaniel, P., Uluagac, A.S.: Curie: policy-based secure data exchange. In: Proceedings of the Ninth ACM Conference on Data and Application Security and Privacy, pp. 121–132. ACM (2019)
15. Centre for Data Ethics and Innovation (CDEI): Privacy enhancing technologies adoption guide (2021). https://cdeiuk.github.io/pets-adoption-guide/
16. Chaudhary, P., Gupta, R., Singh, A., Majumder, P.: Analysis and comparison of various fully homomorphic encryption techniques. In: 2019 International Conference on Computing, Power and Communication Technologies, pp. 58–62 (2019)
17. Clarke, R.: Business cases for privacy-enhancing technologies. In: Computer Security, Privacy and Politics, New York, USA. IRM Press (2008)

18. Cofta, P.: Trust-enhancing technologies. In: Trust, Complexity and Control, West Sussex, England, UK, pp. 187–205. Wiley (2007)
19. Cuno, S., Bruns, L., Tcholtchev, N., Lämmel, P., Schieferdecker, I.: Data governance and sovereignty in urban data spaces based on standardized ICT reference architectures. Data 4(1), 1–24 (2019). https://doi.org/10.3390/data4010016
20. Danezis, G., et al.: Privacy and data protection by design - from policy to engineering (2014). https://doi.org/10.48550/ARXIV.1501.03716
21. Demchenko, Y., de Laat, C., Los, W.: Data as economic goods: definitions, properties, challenges, enabling technologies for future data markets. ITU J. ICT Discov. 1(2), 1–10 (2018). https://doi.org/10.5281/zenodo.2483185
22. Deswarte, Y., Melchor, C.: Current and future privacy enhancing technologies for the internet. Annales des Télécommunications 61, 399–417 (2006). https://doi.org/10.1007/BF03219914
23. Diffie, W., Hellman, M.: New directions in cryptography. IEEE Trans. Inf. Theory 22(6), 644–654 (1976)
24. Dinev, T., Xu, H., Smith, J.H., Hart, P.: Information privacy and correlates: an empirical attempt to bridge and distinguish privacy-related concepts. Eur. J. Inf. Syst. 22(3), 295–316 (2013). https://doi.org/10.1057/ejis.2012.23
25. Dingledine, R., Mathewson, N., Syverson, P.: Tor: the second-generation onion router. In: Proceedings of the 13th USENIX Security Symposium, pp. 1–17 (2004)
26. Dwork, C., McSherry, F., Nissim, K., Smith, A.: Calibrating noise to sensitivity in private data analysis. In: Halevi, S., Rabin, T. (eds.) TCC 2006. LNCS, vol. 3876, pp. 265–284. Springer, Heidelberg (2006). https://doi.org/10.1007/11681878_14
27. Fischer-Hbner, S., Berthold, S.: Privacy-enhancing technologies. In: Computer and Information Security Handbook, 3rd edn, pp. 759–778. Morgan Kaufmann (2017)
28. Gerl, A., Meier, B.: Privacy in the future of integrated health care services-are privacy languages the key? In: 2019 International Conference on Wireless and Mobile Computing, Networking and Communications, pp. 312–317. IEEE (2019)
29. Gil, G., Arnaiz, A., Diez, F.J., Higuero, M.V.: Evaluation methodology for distributed data usage control solutions. In: 2020 Global Internet of Things Summit, Dublin, Ireland, pp. 1–6. IEEE (2020)
30. Goldberg, I., Wagner, D., Brewer, E.: Privacy-enhancing technologies for the internet. In: Proceedings IEEE COMPCON 1997, pp. 103–109. IEEE (1997)
31. Goldreich, O., Oren, Y.: Definitions and properties of zero-knowledge proof systems. J. Cryptol. 7(1), 1–32 (1994). https://doi.org/10.1007/BF00195207
32. Goldwasser, S., Micali, S., Rackoff, C.: The knowledge complexity of interactive proof-systems. In: Proceedings of the Seventeenth Annual ACM Symposium on Theory of Computing, Rhode Island, USA, pp. 291–304. ACM (1985)
33. Grünewald, E., Pallas, F.: TILT: a GDPR-aligned transparency information language and toolkit for practical privacy engineering. In: Proceedings of the 2021 ACM Conference on Fairness, Accountability, and Transparency, Virtual Event, Canada, pp. 636–646. ACM (2021). https://doi.org/10.1145/3442188.3445925
34. Gürses, S., Troncoso, C., Diaz, C.: Engineering privacy by design. In: Conference on Computers, Privacy & Data Protection. CPDP Conferences, pp. 1–21 (2011)
35. Hes, R., Borking, J.J. (eds.): Privacy-Enhancing Technologies: The Path to Anonymity, Revised edn. Registratiekamer, The Hagu (1998)
36. Heurix, J., Zimmermann, P., Neubauer, T., Fenz, S.: A taxonomy for privacy enhancing technologies. Comput. Secur. 53, 1–17 (2015)
37. Hynes, N., Dao, D., Yan, D., Cheng, R., Song, D.: A demonstration of sterling: a privacy-preserving data marketplace. Proc. VLDB Endow. 11(12), 2086–2089 (2018). https://doi.org/10.14778/3229863.3236266

38. Jaatun, M., Tøndel, I.A., Bernsmed, K., Nyre, Å.: Privacy enhancing technologies for information control. In: Privacy Protection Measures and Technologies in Business Organizations: Aspects and Standards, pp. 1–31. IGI Global (2012)

39. Javed, I.T., Alharbi, F., Margaria, T., Crespi, N., Qureshi, K.N.: PETchain: a blockchain-based privacy enhancing technology. IEEE Access Pract. Innov. Open Solutions **9**, 41129–41143 (2021)

40. Jøsang, A., Pope, S.: User centric identity management. In: Proceedings of AusCERT 2005, Brisbane, Australia, pp. 1–13. AusCERT (2005)

41. Kaaniche, N., Laurent, M., Belguith, S.: Privacy enhancing technologies for solving the privacy-personalization paradox: taxonomy and survey. J. Netw. Comput. Appl. **171**, 1–32 (2020)

42. Konečný, J., McMahan, B., Ramage, D.: Federated optimization: distributed optimization beyond the datacenter (2015)

43. Lablans, M., Kadioglu, D., Muscholl, M., Ückert, F.: Exploiting distributed, heterogeneous and sensitive data stocks while maintaining the owner's data sovereignty. Methods Inf. Med. **54**(04), 346–352 (2015)

44. Lauf, F., et al.: Linking data sovereignty and data economy: arising areas of tension. In: Wirtschaftsinformatik 2022 Proceedings, pp. 1–18. AIS (2022)

45. Lee, A.R., Kim, M.G., Won, K.J., Kim, I.K., Lee, E.: Coded Dynamic Consent framework using blockchain for healthcare information exchange. In: 2020 IEEE International Conference on Bioinformatics and Biomedicine, pp. 1047–1050 (2020)

46. Li, T., Sahu, A.K., Talwalkar, A., Smith, V.: Federated learning: challenges, methods, and future directions. IEEE Sig. Process. Mag. **37**(3), 50–60 (2020)

47. López, D., Farooq, B.: A multi-layered blockchain framework for smart mobility data-markets. Transp. Res. Part C Emerg. Technol. **111**, 588–615 (2020). https://doi.org/10.1016/j.trc.2020.01.002

48. Meuser, M., Nagel, U.: The expert interview and changes in knowledge production. In: Interviewing experts, UK, London, pp. 17–42. Palgrave Macmillan (2009)

49. Mühle, A., Grüner, A., Gayvoronskaya, T., Meinel, C.: A survey on essential components of a self-sovereign identity. Comput. Sci. Rev. **30**, 80–86 (2018)

50. Morse, J.M.: Approaches to qualitative-quantitative methodological triangulation. Nurs. Res. **40**(2), 120–123 (1991)

51. Munilla Garrido, G., Schmidt, K., Harth-Kitzerow, C., Luckow, A., Matthes, F.: Exploring privacy-enhancing technologies in the automotive value chain. In: 2021 IEEE International Conference on Big Data, Orlando, USA, pp. 1–8. IEEE (2021)

52. Munilla Garrido, G., Sedlmeir, J., Uludağ, Ö., Alaoui, I.S., Luckow, A., Matthes, F.: Revealing the landscape of privacy-enhancing technologies in the context of data markets for the IoT: a systematic literature review (2021)

53. Nast, M., et al.: Work-in-progress: towards an international data spaces connector for the Internet of Things. In: 2020 16th IEEE International Conference on Factory Communication Systems, Porto, Portugal, pp. 1–4. IEEE (2020)

54. Niu, C., Zheng, Z., Wu, F., Gao, X., Chen, G.: Achieving data truthfulness and privacy preservation in data markets. IEEE Trans. Knowl. Data Eng. **31**(1), 105–119 (2019). https://doi.org/10.1109/TKDE.2018.2822727

55. OMTP: Advanced trusted environment. Technical report, OMTP Limited (2009)

56. Otto, B., et al.: Reference architecture model for the industrial data space. Technical report, Fraunhofer Gesellschaft (2017)

57. Otto, B., Jarke, M.: Designing a multi-sided data platform: findings from the International Data Spaces case. Electron. Mark. **29**(4), 561–580 (2019). https://doi.org/10.1007/s12525-019-00362-x

58. Palmieri, P., Pouwelse, J.: Key management for onion routing in a true peer to peer setting. In: Yoshida, M., Mouri, K. (eds.) IWSEC 2014. LNCS, vol. 8639, pp. 62–71. Springer, Cham (2014). https://doi.org/10.1007/978-3-319-09843-2_5

59. van Panhuis, W.G., et al.: A systematic review of barriers to data sharing in public health. BMC Public Health **14** (2014). https://doi.org/10.1186/1471-2458-14-1144

60. Papadopoulos, P., Abramson, W., Hall, A.J., Pitropakis, N., Buchanan, W.J.: Privacy and trust redefined in federated machine learning. Mach. Learn. Knowl. Extract. **3**(2), 333–356 (2021)

61. Parra-Arnau, J., Rebollo-Monedero, D., Forné, J.: Privacy-enhancing technologies and metrics in personalized information systems. In: Navarro-Arribas, G., Torra, V. (eds.) Advanced Research in Data Privacy. SCI, vol. 567, pp. 423–442. Springer, Cham (2015). https://doi.org/10.1007/978-3-319-09885-2_23

62. Pennekamp, J., et al.: Dataflow challenges in an internet of production: a security & privacy perspective. In: Proceedings of the ACM Workshop on Cyber-Physical Systems Security & Privacy, London, UK, pp. 27–38. ACM (2019)

63. Runeson, P., Höst, M.: Guidelines for conducting and reporting case study research in software engineering. Empir. Softw. Eng. **14**(2), 131–164 (2009). https://doi.org/10.1007/s10664-008-9102-8

64. Samarati, P.: Protecting respondents identities in microdata release. IEEE Trans. Knowl. Data Eng. **13**(6), 1010–1027 (2001). https://doi.org/10.1109/69.971193

65. Samarati, P., Sweeney, L.: Protecting privacy when disclosing information: k-anonymity and its enforcement through generalization and suppression. Technical report, Data Privacy Lab (1998)

66. Sarabia-Jacome, D., Lacalle, I., Palau, C.E., Esteve, M.: Enabling industrial data space architecture for seaport scenario. In: 2019 IEEE 5th World Forum on Internet of Things, Limerick, Ireland, pp. 101–106. IEEE (2019)

67. Schmidt, K., Ullrich, A., Eigelshoven, F.: From exploitative structures towards data subject-inclusive personal data markets - a systematic literature review. In: Proceedings of the 29th European Conference on Information Systems (2021)

68. Seničar, V., Jerman-Blažič, B., Klobučar, T.: Privacy-enhancing technologies-approaches and development. Comput. Stand. Interfaces **25**(2), 147–158 (2003). https://doi.org/10.1016/S0920-5489(03)00003-5

69. Shamir, A.: How to share a secret. Commun. ACM **22**(11), 612–613 (1979). https://doi.org/10.1145/359168.359176

70. Sharma, S., Chen, K., Sheth, A.: Toward practical privacy-preserving analytics for IoT and cloud-based healthcare systems. IEEE Internet Comput. **22**(2), 42–51 (2018). https://doi.org/10.1109/MIC.2018.112102519

71. Song, D.X., Wagner, D., Perrig, A.: Practical techniques for searches on encrypted data. In: Proceeding 2000 IEEE Symposium on Security and Privacy, Berkeley, USA, pp. 44–55. IEEE (2000). https://doi.org/10.1109/SECPRI.2000.848445

72. Spiekermann, S., Novotny, A.: A vision for global privacy bridges: technical and legal measures for international data markets. Comput. Law Secur. Rev. Int. J. Technol. Law Pract. **31**(2), 181–200 (2015)

73. Sweeney, L.: k-anonymity: a model for protecting privacy. Internat. J. Uncertain. Fuzziness Knowl.-Based Syst. **10**(05), 557–570 (2002)

74. Syverson, P., Goldschlag, D., Reed, M.: Anonymous connections and onion routing. In: Proceedings of the 1997 IEEE Symposium on Security and Privacy, pp. 44–54 (1997)

75. Thurmond, V.A.: The point of triangulation. J. Nurs. Scholarsh. **33**(3), 253–258 (2001). https://doi.org/10.1111/j.1547-5069.2001.00253.x

76. Trask, A., Bluemke, E., Garfinkel, B., Cuervas-Mons, C.G., Dafoe, A.: Beyond privacy trade-offs with structured transparency (2020)
77. Vom Brocke, J., Simons, A., Niehaves, B., Riemer, K., Plattfaut, R., Cleven, A.: Reconstructing the giant: on the importance of rigour in documenting the literature search process. In: Proceedings of the 17th European Conference on Information Systems, Verona, Italy, pp. 1–12. AIS (2009)
78. Webster, J., Watson, R.T.: Analyzing the past to prepare for the future: writing a literature review. MIS Q. **26**(2), xiii–xxiii (2002)
79. Yao, A.C.: Protocols for secure computations. In: 23rd Annual Symposium on Foundations of Computer Science, Chicago, IL, USA, pp. 160–164. IEEE (1982)
80. Zöll, A., Olt, C.M., Buxmann, P.: Privacy-sensitive business models: barriers of organizational adoption of privacy-enhancing technologies. In: Proceedings of the 29th European Conference on Information Systems, pp. 1–21. AIS (2021)
81. Zrenner, J., Möller, F.O., Jung, C., Eitel, A., Otto, B.: Usage control architecture options for data sovereignty in business ecosystems. J. Enterp. Inf. Manag. **3**(32), 477–495 (2019)

Threat Detection and Mitigation with Honeypots: A Modular Approach for IoT

Simão Silva[1]([⊠]) [ID], Patrícia R. Sousa[1,2] [ID], João S. Resende[3] [ID],
and Luís Antunes[2] [ID]

[1] CRACS - INESCTEC, Porto, Portugal
simao.sfos@gmail.com, psousa@dcc.fc.up.pt
[2] University of Porto, Porto, Portugal
lfa@dcc.fc.up.pt
[3] NOVA-LINCS, Faculdade de Ciências e Tecnologia, Universidade NOVA de Lisboa,
Lisbon, Portugal
jresende@fct.unl.pt

Abstract. A honeypot is a controlled and secure environment to examine different threats and understand attack patterns. Due to the highly dynamic environments, the growing adoption and use of Internet of Things (IoT) devices make configuring honeypots complex. One of the current literature challenges is the need for a honeypot not to be detected by attackers, namely due to the delays that are required to make requests to external and remote servers. This work focuses on deploying honeypots virtually on IOT devices. With this technology, we can use endpoints to send specific honeypots on recent known vulnerabilities on IOT devices to find and notify attacks within the network, as much of this information is verified and made freely available by government entities. Unlike other approaches, the idea is not to have a fixed honeypot but a set of devices that can be used at any time as a honeypot (adapted to the latest threat) to test the network for a possible problem and then report to Threat Sharing Platform (TSP).

Keywords: CVE · Honeypot · Internet of Things · Intrusion detection · Security · Vulnerability

1 Introduction

The evolution of technology is changing society and the environment around us. With users spending more time looking at a screen than sleeping [10], it is important to alert users to the hardships they may face online regarding security and privacy. As the software is not unbreakable and new vulnerabilities are identified and disclosed daily, devices and data can be compromised. New vulnerabilities are threats that an attacker can use to take advantage of the user, creating severe consequences.

S. Katsikas and S. Furnell (Eds.): TrustBus 2022, LNCS 13582, pp. 66–80, 2022.
https://doi.org/10.1007/978-3-031-17926-6_5

The growing number of cyberattacks and their extensive damage has changed the mindset of organizations, with more resources being applied to cyber security. According to statistics from AV-TEST as of January 2021 [3], the security community knew over a billion malicious executable scripts. Digital transformation has also meant increased cybercrime, often associated with significant financial losses for individuals and organizations. As new daily vulnerabilities are identified and disclosed, attackers constantly find new ways to compromise devices. However, the increase in the number of cyberattacks and their extensive damage has changed organizations' mindset, applying more resources, specific teams, and dedicated tools to deal with this problem.

The use of honeypots has been a vital tool to face the increase in cyberattacks. A honeypot is configured to detect, circumvent, or prevent unauthorized use of information systems. A honeypot is a controlled and secure environment to examine different threats and understand attack patterns. It is possible to analyze attack traffic independently of regular network traffic and away from critical infrastructure. Thus, security teams can focus on analyzing just the threat. Honeypots are versatile and are not addressed exclusively to a specific problem like other standard solutions (firewalls and antivirus). Given its functioning, it becomes an essential information tool to spot existing threats and possibly new ones, which makes them an important asset, especially relevant for Security Operations Center (SOC) teams since they possess a significant database of attacks' artifacts and, consequently, a significant source of threat intelligence. Honeypot systems are becoming a mandatory component for cybersecurity defense by working with threat intelligence platforms.

Although many honeypots are aimed at different purposes, these systems have some constraints. Given the diversity of devices, there are still limitations in using these systems outside of x86 architectures, leaving out capable devices such as Raspberry Pi, which align computational power with low energy consumption. Another issue is regarding its use. With the proliferation of the cloud, we are witnessing the phenomenon of services migration to external servers, which, in the case of honeypots, the trend is to use remote machines as baits, and often, the traffic is redirected from the local to the cloud network. In this case, resorting to this method allows the attacker to quickly detect that it is in a honeypot by measuring the latency times of Internet Control Message Protocol (ICMP) ECHO requests, thus making the honeypot ineffective.

This paper contributes to the existing literature with the creation of an automatic honeypot instance deployment when a vulnerability is present in a device in the network, capable of running a variety of devices, including low-power, performance-constraint devices, and provide a mechanism that can discretely monitor honeypots instances to collect intelligence of tactics and techniques of intruders. Also, overall, we contribute with an autonomous threat detection flow able to analyze the diversity of devices in the network and detect and mitigate, if possible, its vulnerabilities.

2 Related Work

In IOT-based networks, new devices entering the network are automatically configured due to their open nature, which leaves these networks subject to many attacks, as described in [18]. Given these systems' complexity and configurations, a common approach is to use low-interaction honeypots as they are easier to install and configure and with low risk. *Dshield* [1], *Glastopf* [26] and *Open-Canary* [4] can monitor requests and alert of potential unwanted traffic but, due to its characteristics, has limitations on the information gather from attackers which unable the trace of attacks. Also, the attackers can easily identify them due to their appearance and behavior [11,20].

Honeyd [21] is a simulated honeypot environment intended to simulate the virtual network topology to filter network packets based on user preferences. When sending a response packet, the personality engine makes it match the network behavior of the configured operating system personality [22]. As it can emulate operating systems, *Honeyd* can appear to the attacker as a router, web server, or DNS server, allowing the honeypot to blend into existing networks. With this behavior, *Honeyd* can spoof responses about active fingerprint measurements, such as those used by the NMAP tool.

The increasing need for honeypots has led to the concept of Honeypot-as-a-Service (HaaS), where, instead of being configured locally, they are made available to users by cloud providers, thus eliminating the need to worry about hardware, software, and human resources to create and maintain the honeypot. There are public cloud provider honeypots [16] that, although it reduces costs, still require labor to apply and maintain the configurations. An evaluation of this solution [14] shows that it still involves human resources to use and maintain the settings, and the traffic flow is in order.

Honeyd's low-interaction honeypots [6] explore the idea of honeypots that adapt to the needs and changes of organizations and use unassigned IP addresses to launch instances. With traffic being diverted in the background from one honeypot to another, *Honeyd* honeypots are more reliable for intruders. However, this requires configuring physical machines dedicated exclusively to the honeypot network, and as we would need to mimic more rogue systems, more IP addresses also need to be available.

Regarding the industrial sector, there is also a concern about safety, namely for Programmable Logic Controllers (PLCs). *HoneyPLC* [19] overcomes the limitations of current honeypot implementations for PLCs, with easy fingerprinting and low interaction levels being some examples. However, the system remains stagnant and unable to adapt to new changes in existent threats.

A comprehensive solution for honeypot systems focused on IOT environments is the YAKSHA [17] project, which consists of the development of a cloud-based platform that allows an organization to perform continuous monitoring and penetration tests of its infrastructure without the need of need to maintain such infrastructure nor have dedicated staff. It provides a way to create custom honeypots tailored to specific needs that include installing a specific set of services that will be used to lure attackers. Although it is designed for the IOT environment, the honeypot deployment, like the malware analysis tool, requires additional hardware based on the x86 architecture. Also, while a mechanism for patching and configuring IDS is mentioned in vulnerability detection, it is unclear how this process is accomplished.

Table 1 shows the comparison of characteristics of current state-of-the-art against our proposed solution (Sandboxing for Threat Detection and Mitigation (STDM)).

Table 1. State of the art's features summary

	[1]	[26]	[4]	[21]	[16]	[14]	[19]	[17]	STDM
Honeypot deployment	X	X	X	X	X	X	X	X	X
Device profiling				X	X	X	X	X	
Honeypot customization				X	X	X	X	X	X
Decentralized deployment supported	X		X					X	X
CVE vulnerability scanning									X
IDS connection								X	X
TSP connection									X
ARM compatibility	X	X	X						X

All these solutions lack the characteristics of environmental independence and mobility. The presented honeypots cannot be deployed on any device, at any time, regardless of its architecture. Given the highly dynamic environments associated with IOT, honeypots must also be interchangeable between available devices on the network, as well as the ability to be easily replicated across different devices.

3   Architecture

Gathering threat knowledge is one of the most important and necessary tasks in today's cybersecurity. The increase in attacks led to changes in collecting knowledge of (potential) threats and how to take advantage of that knowledge.

Figure 1 shows the overview of the architecture of our system. The Threat Sharing Platform (TSP) component will update their database with the Common Vulnerabilities and Exposures (CVE) entries (1). The *STDM Central Coordinator* will act as a go-between managing the device scanning history and honeypots. From time to time, the devices will send their list of software installed to the *STDM Central Coordinator* that, in turn, requests the information about CVEs from the TSP (2). For each CVE entry, the coordinator parses the Common Platform Enumeration (CPE) list - a standard that specifies a structured naming scheme for software based on the uniform resource identifiers syntax - and searches for a match in the list received from the devices (3–4). When a match is found, the *STDM Central Coordinator* will select a random device and launch a honeypot with the same operating system and software version of the match (5). Inside the honeypot, an admin user can verify if the match is not a false positive and provide that information to the *STDM Central Coordinator* (6). From the honeypot logs (7), an administrator user can observe the malicious requests and can create, if possible, rules to be applied to the *Host-based Intrusion Detection System (HIDS)* of all devices (8).

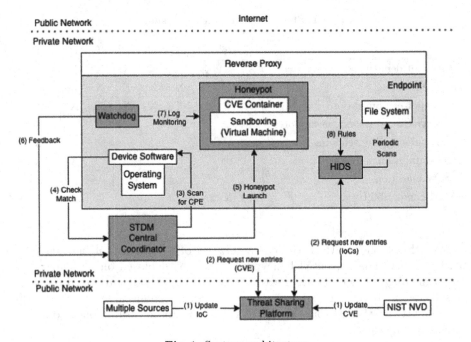

Fig. 1. System architecture

In the following subsections, we will describe the highlighted components of our architecture.

3.1 Threat Sharing Platform

This component is responsible to store information from Indicators of Compromise (IoC) and CVE. From time to time, the platform will retrieve IoC from public and trustworthy feeds and act as a local mirror database of *National Institute of Standards and Technology (NIST)*'s National Vulnerability Database (NVD) database of CVE entries.

Even in security breaches, we can still learn with the shreds of evidence left by the intruder. IoC [9] enhances the honeypot by capturing malicious activities on the network at their initial stage, preventing them from becoming more significant problems and compromising the security of the infrastructures.

Also, in our approach, we intend to enrich our knowledge through known public vulnerabilities, namely CVEs [12]. As they are public, we can recreate the environments that mimic that CVE and use it to gather IoC that can be later used to enhance our HIDS, as shown in steps 5, 7, and 8 of Fig. 1. Those environments are launched as honeypots and are exposed to the Internet as decoy services.

3.2 Host Intrusion Detection System

We use the component of HIDS to detect and automate the honeypot initialization process according to what is happening in the network (environment).

To improve security on our devices, we have deployed two HIDS systems - one that offers anomaly and signature variants and one that offers Intrusion Prevention System (IPS) capabilities - on each device, as these systems are designed to increase protection against internal and external threats. In addition, they can monitor network traffic to and from the machine, observe running processes, and inspect system logs for patterns that match known cyberattacks. While they can be seen as limited due to their low visibility (limited to the host, which decreases the decision-making context), their deep visibility into the internals of the host allows the security teams to analyze activities in a high level of detail. Thus, unlike network-based intrusion detection systems [15], they can directly access and monitor data files and processes of the targeted system.

3.3 STDM Central Coordinator

This component is the gateway between the threat-sharing platform and the devices in the network. It is responsible for checking if a device in our network has vulnerable software, keeping records of those threats, and launching the honeypots containing them.

3.4 Watchdog

Watchdog is an agent installed on the host system to allow users to access any honeypot file without having to directly interact with it or learn the back-end command syntax to access it. From *Watchdog*, a user can monitor the logs of the vulnerable service running on the honeypot instance. With this information, it can analyze logs and analyze malicious queries that can later be manually added to the IoC database in TSP, which allows the platform to be aware of insider attacks viewed or received by honeypots, allowing to prioritize threats in a global overview.

The *Watchdog* can be adapted to any installation of service in the virtual machine.

3.5 Honeypot

The *honeypot* instances are virtual machines with service(s) exposed to the Internet. Those services are installed directly on the file system or using containers (to avoid external changes that can interfere with the service availability and expected behavior). The vulnerable services are represented by the *CVE container* inside the *Sandboxing Virtual Machine* in the *Honeypot* (in the Fig. 1). The instances are prepared to run in x86-64 and ARM architectures, thus enabling (virtually) the launching of an instance on any device on the network. Also, this mobility enables a given instance running a given service can be easily shifted between devices.

These instances require isolation techniques so that the attacker can not gain access to the host system and damage it. The typical strategy used in the cybersecurity world is sandboxing, which is a security mechanism that tricks an application or program into thinking it is running on a regular computer [5]. It allows us to provide services to intruders in a tightly controlled environment without allowing the services and the intruders' actions to harm the host device.

4   Implementation

Following the presentation of our architecture and its components in Sect. 3, this chapter describes the technical details of the implementation of our system, the decisions made, and problems that emerged during said implementation.

4.1   Threat Sharing Platform

For our architecture, we opt to use Malware Information Sharing Platform (MISP) as our threat (intelligence) sharing platform [27]. Interactions with MISP can be done by its Representational State Transfer (REST) interface. Using PyMISP [23], we can manage the platform and add new functionalities. In our case, we use it to keep our database of IoC and CVEs updated.

Regarding CVEs, we keep our database updated by a custom module-like script we developed, henceforth referred to as *CVE module*. That module will use NIST's REST service to periodically request new or recent modified entries from the CVE database since the last request. The service will answer with an empty response - meaning that we are up to date - or a list of entries with their respective details in JavaScript Object Notation (JSON) format.

For each element in that list, we must verify the contents of the *input* field. If the field is empty, the CVE is under analysis, and the information about this vulnerability is still preliminary. If not, we make a subsequent request to retrieve the CPE software list affected by the vulnerability. Using the information from both requests, namely ID, description, reference links, and CPE list, we use the CVE module to create a MISP event with all these attributes and store it on the platform, making it available to the other components of our system.

4.2   STDM

This component is the bridge between the network devices and the TSP. The component comprises a server with a database storing the values of the device's last authentication timestamp (for debugging) and several vulnerabilities found (for statistical purposes). All communications between devices and our server use SSL/TLS to provide confidentially and integrity. The devices will authenticate themselves into the component using a certificate-based authentication where each device has its own issued certificate that was later manually installed. This method allowed us easier management of the devices, given that only devices with a valid certificate installed can access our system. Then, they will periodically send the names and versions of the software installed on the host to the server. In turn, the server will request the TSP platform for the list of CVEs and respective CPEs and verify if there is a match against the information received from the device. If a match is found, it increments the positive matches statistics, launches a honeypot instance with the same operating system, and informs the admin of the vulnerable software version to install it on the honeypot later. The honeypot instance starts on a random device with the necessary resources to run the instance and is available in the network at the launch time.

The software matching is not linear and requires a pre-processing step. Given that the names in NVD's database are in CPE syntax and differ from the syntax used by operating systems, we attempt to translate the CPE syntax closer to the names on the operating systems. However, we notice a high rate of false positives in some situations, which we attribute to the versioning naming format. In these cases, a warning is displayed to an admin user that has to check if it is a match manually.

4.3   Host Intrusion Detection System

For our solution, we opt to use a combined solution of HIDS by leveraging the Fail2ban's IPS characteristics. As HIDS are *passive* in its essence, meaning they identify but do not prevent suspicious activity, we chose to add an IPS system as they are active in preventing those suspicious activities. Thus, to take advantage of all of their features, we used OSSEC (HIDS) combined with Fail2ban (HIDS/IPS).

4.4   Isolated Environments

Honeypot's security creates a vulnerability in the host that is deliberately singled out for attackers to use. In addition, deploying these systems in secure environments is necessary so that activities on honeypot systems do not affect the host. This section describes the technologies and methods applied to achieve this goal with that goal in mind.

Virtualization was used because the kernel and libraries are not shared between host and guest to prevent an attacker from accessing the host machine. From the functionality point of view, it allows the creation of several isolated environments from single hardware, thus allowing better usage of resources. While containers are highly supported for our architectures (x86 and ARM), the isolation provided by virtual machines was the critical factor to consider.

Virtualization technology has many solutions for x86 architectures, but the same does not apply to devices with ARM architectures. However, continuous improvements in boards, such as the Raspberry Pi, have made it feasible to use Kernel-based Virtual Machine (KVM), a Linux kernel module that allows the Linux kernel to act as a hypervisor, which has become more accessible for users with Linux distributions with the pre-compiled Linux KVM kernel module.

The chosen virtual machine provider was Multipass [8] due to the support of different architectures and operating systems, allowing it to scale its use to any system. However, to manage virtualization, we had to change Multipass settings. By default, it uses the QEMU hypervisor [7] that, although it runs perfectly on x86-based processors, presented some incompatibility issues when running on ARM processors, making Multipass unable to run. So, we change the hypervisor to LXD [2], which is a REST API that connects to *libxlc*, which is the Linux Containers (LXC) library - a solution for virtualizing software at the operating system level within the Linux kernel [24] that can monitor the virtual machines from the host's filesystem.

The honeypot system is a Multipass virtual machine deployed by the *STDM Central Coordinator* component. Currently, only Ubuntu cloud images can be deployed automatically. However, the usage of custom images is also supported.

The system's image is very similar to a non-graphical version of Ubuntu Server, with no visible distinguishability that can hint to the intruder that it is a decoy. Once up and running, an admin user sets up the vulnerable service, including the additional required software and the necessary port-forwarding with IPtables. The services are installed preferably from the source code of the corresponding vulnerable version and, when possible, deployed on containers to take advantage of its portability so that installation is environment independent and the service easily deployed (it can start in just a few seconds) with minimal overhead.

4.5   Watchdog

We have two ways to access the honeypot: the Multipass interface or the LXC/LXD interface. Both options allow us to access the honeypot and monitor the services' logs, but they can reveal to the attacker that he/she is in a decoy system. Using the Multipass interface, an attacker can monitor the SSH log file and discover a user with *Sudo* privileges monitoring the log files, which can be an admin user and thus jeopardize the decoy system. Instead, if the LXC/LXD interface is used, the attacker can be suspicious of the system if it discovers that the *lxd-agent* is running (given this is necessary for the host-honeypot communication) or if a process monitoring some log file own by the root is running. The resolution of these issues was two: the first one is mounting the */proc* directory with *hidepid=2* flag, thus denying users access to processes besides theirs; the second one is by unmounting and mounting the system image file (.img) used to support the virtual machine on the host's file system. This second option comes with the inconvenience of repeating the unmount/mount routine due to a limitation on the image file that supports the virtual machine file system. When mounted, we notice that later changes made within the virtual machine's file system are not reflected in the mount point and vice versa. However, we can see earlier changes if we repeat the unmount/mount routine on the image file. This limitation forces us to redo the unmount/mount routine to get new data from the log files, making real-time log monitoring impossible.

Given the options above, we opt to use the LXC/LXD interface. Even though the flag usage can indicate the presence of a decoy system, it can also be seen as a default security measure employed by an admin. Also, considering the usability trade-off between the two resolutions, the usage of the LXC/LXD interface provides a better user interaction.

To avoid the need for admin users to learn and remember the syntax of the command to access the instance, we developed a script that abstracts the backend command and accepts user input to execute commands as if the users were inside the instance.

5   Evaluation

In this section, we aim to evaluate the performance of our system. Our evaluation focuses on the measurement of delay between the time attack is detected (i.e., the malicious request is detected) on the honeypot and the time the central coordinator receives that information.

The goal of our evaluation is to demonstrate the feasibility of the solution. For the performance evaluation, the main challenge was to measure and compare with a standard service in the literature. An example of this service type is Apache2 [13], which is one of the most used web servers. We use a vulnerability of this service to demonstrate our prototype (CVE-2019-10092) using the Apache HTTPd 2.4.38 server (this issue affects the versions from 2.4.7 up to 2.4.51, included).

5.1   Environment Description

For our scenario, we used a Raspberry Pi 4 Model B with 4 GB of RAM with a Broadcom BCM2711 1.5 GHz Quad-Core 64-bits processor connected through an Ethernet cable to our router Thomson TG784n. Also, we acquired a virtual machine from Microsoft Azure with 2 GB of RAM and an Intel Xeon E5-2673 v3 2.40 GHz processor truncated at two cores. Both used Ubuntu 20.04 LTS as the operating system.

5.2   Methodology

We designed our testing scenarios to consider different honeypot installations. We tested our system in the following scenarios: Setup 1 (HaaS with physical machines): cloud virtual machine exclusively dedicated to being a honeypot; Setup 2 (HaaS with virtual machines and using LXD): using Multipass instances as honeypots on the cloud virtual machine to evaluate the impact of virtualization overhead; and Setup 3 (*STDM*): our solution runs in a local network with a Raspberry Pi running Multipass instances as honeypots.

The tests were performed to explore the differences against the HAAS concept and usage of cloud machines to compare the delay times between on-premise and off-premise solutions (State-of-the-Art). Given the usage of virtualization, we also tested its impact on the overall results. Scenarios from Setup 1 and Setup 2 allow us to test and compare the State-of-the-Art solutions with our solution (Setup 3).

Regarding deployment, we launch the vulnerable Apache Server from a container available on a Docker Registry to simplify the installation process in all setups. For setups 1 and 2, to allow queries from cloud machines to reach the central coordinator, we had to configure IPTables rules and delete the default Azure forwarding rules causing the requests to be dropped. For setups 2 and 3, we deployed a Multipass virtual machine, installed the Docker software, and configured the necessary redirects from the host to the virtual machine to reach the server from the outside.

The attack scenario was performed on a client-server basis running an environment to mimic the CVE-2019-10092 in the Apache HTTPd 2.4.38 server.

We measure the delay times between the moment the malicious request is detected and when the information is recognized by the *STDM Central Coordinator*. For this, a specific tool was designed, located precisely in the *STDM Central Coordinator*, which provides the network latency generated (in milliseconds) by our solution. The exploitation was performed using a bash script containing a cURL request to the server's IP address.

5.3   Results

This section aims to show the results and comparison with state-of-the-art.

Table 2 presents the results for each of the three scenarios of network throughput. For gathering the network information, we used *iPerf* [25], which is software used to test the network bandwidth. We collected two-time samples to measure the setup performance only in network bandwidth. The results presented in the table are the mean and standard deviation for the three setups and both measurements (TCP bandwidth and UDP jitter).

Table 2. Latency results for the three setups

	Setup 1	Setup 2	Setup 3
TCP bandwidth (Mbits/sec sd)	1,22 0,89	1,17 1,06	30,29 2,91
UDP jitter (ms sd)	4,02 1,12	4,32 2,88	0 0

Figure 2 represents the delay associated with each implementation scenario. The latency of the process between detecting the malicious request in the honeypot instance and the moment when the *STDM Central Coordinator* receives this information. To do this, we collect three samples for each number of requests.

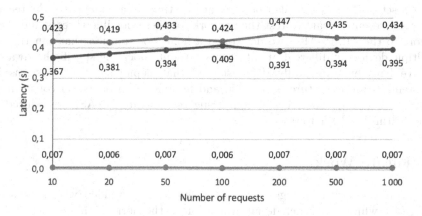

Fig. 2. Latency (ms) per number of requests

The results of Setup 1 (HAAS with physical machines) show a delay in the order of 390 ms with minimal oscillations. Then, Setup 2 (HAAS with virtual machines and using LXD) presents small delay increments compared to Setup 1, showing that virtual machines have a residual overhead on performance and is, therefore, a good choice when looking for multiple deployments and host isolation. On the other hand, Setup 3 (STDM - our implementation) shows the main advantage of having Honeypots on the network endpoints because the delay associated with this task is reduced to close to 0 with on-premises honeypots. Similar behavior could be achieved with Honeypots running on a separate Local Area Network (LAN), but these solutions raise concerns about the actual usability of the honeypot network (as shown in related works).

6   Conclusion

In this work, we studied and created a solution to autonomously detect and prevent threats based on public information of known treats and specialized software that gathers, parses, and analyzes information.

We offer a new security approach to deploying dynamic local honeypots capable of running on devices not previously used for this purpose. As far as we know, this is the first proposal for a dynamic honeypot that can be installed on any network device in order to test network devices, being capable of confusing the attacker, as he/she never knows if it is a network device or a honeypot placed for this purpose. IOT environments benefit from this solution as sensors are often the most vulnerable devices on a network.

It combines IDS and IPS systems to detect and automate the honeypot initialization process according to what is happening in the network (environment).

This process can be used as the first line of defense for active or passive intrusion detection and prevention, as it also allows insights into new attacks.

Contrary to previous work, the idea is not to have a fixed honeypot but to have a set of devices that can be used at any time as a honeypot (adapted to the most recent threat) to test the network for these possible threats and then report to the MISP. Also, we developed a honeypot solution that can run on multiple devices of different architectures without allocating specific hardware.

We have experimentally demonstrated that deploying decoy systems in on-premises infrastructure is possible and feasible, even on power-constrained devices. Regarding latency and comparing to standard HAAS, we reduced it from 360 ms to closer to 0 ms.

7   Future Work

As future research challenges, the paper should be extended with real-world use cases, which must include an evaluation of the normal functioning of IoT devices while acting as a honeypot and other evaluation parameters such as memory consumption and power consumption in a real-world IoT use case.

We think another exciting path focuses on enhancing the system with human-in-the-loop mechanisms. In addition to the supervision performed automatically by the system, a user must be able to add their domain knowledge that could identify points of failure and new rules for improving the detection of abnormal patterns, reducing false alarm rates and adversarial attacks. As misclassification can have serious consequences, human-in-the-loop should be used to confirm all patterns and decisions. In this sense, human-in-the-loop in IPS and IDS systems can validate the abnormal events and generate rules to feed the first defense layer. On our current implementation, an administrator can add rules with Fail2ban, but it is essential to make more steps for the overall system, allowing the launch of honeypots with new rules.

If an attacker gains access to the system, they may attack the system (as explained in the security analysis section), but one solution to blocking this type of attack is to use multiple software vendors to perform this process. In combination with Byzantine Fault Tolerance (BFT), it will mitigate and detect another vector of attacks on the host machine, allowing to identify wrong answers from the system.

Acknowledgements. This work is financed by National Funds through the Portuguese funding agency, FCT - Fundação para a Ciência e a Tecnologia, within project LA/P/0063/2020.

The work of Simão Silva was partially funded by the SafeCities POCI-01-0247-FEDER-041435 project through COMPETE 2020 program. The work of Patrícia R. Sousa was partially supported by the Project "City Catalyst - Catalisador para cidades sustentáveis", with reference POCI-01-0247-FEDER-046119, financed by Fundo Europeu de Desenvolvimento Regional (FEDER), through COMPETE 2020 and Portugal 2020 programs. João S. Resende's work was partially supported by the EU H2020-SU-ICT-03-2018 Project No. 830929 CyberSec4Europe (cybersec4europe.eu).

National Funds also partially supported this work through the Agência para a Modernização Administrativa, program POCI - Programa Operacional Competitividade e Internacionalização, within project POCI-05-5762-FSE-000229.1.

References

1. DShield Honeypot. https://isc.sans.edu/honeypot.html. Accessed 25 Jan 2022
2. LXD - Introduction. https://linuxcontainers.org/lxd/introduction/. Accessed on 26 July 2021
3. Malware Statistics & Trends Report. https://www.av-test.org/en/statistics/malware/. Accessed 26 Jan 2021
4. Opencanary honeypot. https://opencanary.readthedocs.io/en/latest/. Accessed 25 Jan 2022
5. Virtualization-based Sandboxes are vulnerable to advanced malware. https://www.lastline.com/blog/virtualization-based-sandboxes/. Accessed 28 May 2021
6. Artail, H., Safa, H., Sraj, M., Kuwatly, I., Al-Masri, Z.: A hybrid honeypot framework for improving intrusion detection systems in protecting organizational networks. J. Comput. Secur. **25**(4), 274–288 (2006). https://doi.org/10.1016/j.cose.2006.02.009, https://www.sciencedirect.com/science/article/pii/S0167404806000587
7. Bellard, F.: QEMU, a fast and portable dynamic translator. In: USENIX Annual Technical Conference, FREENIX Track. vol. 41, p. 46. California, USA (2005)
8. Canonical: Multipass orchestrates virtual Ubuntu instances (2015). https://github.com/canonical/multipass. Accessed 20 July 2021
9. Catakoglu, O., Balduzzi, M., Balzarotti, D.: Automatic extraction of indicators of compromise for web applications. In: Proceedings of the 25th International Conference on World Wide Web, pp. 333–343 (2016)
10. Editors, I.I.: US adults added 1 hour of digital time in 2020, January 2021. https://www.emarketer.com/content/us-adults-added-1-hour-of-digital-time-2020. Accessed 19 Aug 2021
11. Franco, J., Aris, A., Canberk, B., Uluagac, A.S.: A survey of honeypots and honeynets for internet of things, industrial internet of things, and cyber-physical systems. IEEE Commun. Surv. Tutor. **23**(4), 2351–2383 (2021)
12. Guo, M., Wang, J.A.: An ontology-based approach to model common vulnerabilities and exposures in information security. In: ASEE Southest Section Conference (2009)
13. Hu, Y., Nanda, A., Yang, Q.: Measurement, analysis and performance improvement of the Apache web server. In: 1999 IEEE International Performance, Computing and Communications Conference (Cat. No. 99CH36305), pp. 261–267. IEEE (1999)
14. Jafarian, J., Niakanlahiji, A.: Delivering Honeypots as a Service, January 2020. https://doi.org/10.24251/HICSS.2020.227
15. Javaid, A., Niyaz, Q., Sun, W., Alam, M.: A deep learning approach for network intrusion detection system. EAI Endor. Trans. Secur. Saf. **3**(9), e2 (2016)
16. Khan, N.F., Mohan, M.M.: Honey pot as a service in cloud. Int. J. Pure Appl. Math. **118**(20), 2883–2888 (2018)
17. Kostopoulos, A., et al.: Realising honeypot-as-a-service for smart home solutions. In: 2020 5th South-East Europe Design Automation, Computer Engineering, Computer Networks and Social Media Conference (SEEDA-CECNSM), pp. 1–6. IEEE (2020)

18. La, Q.D., Quek, T.Q., Lee, J., Jin, S., Zhu, H.: Deceptive attack and defense game in honeypot-enabled networks for the internet of things. IEEE Internet Things J. **3**(6), 1025–1035 (2016)

19. López-Morales, E., et al.: Honeyplc: a next-generation honeypot for industrial control systems. In: Proceedings of the 2020 ACM SIGSAC Conference on Computer and Communications Security, pp. 279–291 (2020)

20. Mphago, B., Mpoeleng, D., Masupe, S.: Deception in web application honeypots: case of Glastopf. Int. J. Cyber-Secur. Digit. Foren. **6**(4), 179–185 (2017)

21. Provos, N.: Honeyd - a virtual honeypot daemon. In: 10th DFN-CERT Workshop, Hamburg, Germany. vol. 2, p. 4 (2003)

22. Provos, N., et al.: A Virtual Honeypot Framework. In: USENIX Security Symposium, vol. 173, pp. 1–14, January 2004

23. PyMISP, G.: PyMISP - Python Library to access MISP. PyMISP. https://github.com/MISP/PyMISP. Accessed 20 Feb 2021

24. Senthil Kumaran, S.: Practical LXC and LXD: linux Containers for Virtualization and Orchestration. Springer, Cham (2017). https://doi.org/10.1007/978-1-4842-3024-4

25. Tirumala, A.: IPERF: the TCP/UDP bandwidth measurement tool (1999). http://dast.nlanr.net/Projects/Iperf/

26. Tools, K.Y.: Glastopf - a dynamic, lowinteraction web application honeypot (2010)

27. Wagner, C., Dulaunoy, A., Wagener, G., Iklody, A.: MISP: The design and implementation of a collaborative threat intelligence sharing platform. In: Proceedings of the 2016 ACM on Workshop on Information Sharing and Collaborative Security, pp. 49–56 (2016)

Homomorphic Encryption
in Manufacturing Compliance Checks

Aikaterini Triakosia[1]($\boxtimes$), Panagiotis Rizomiliotis[1], Konstantinos Tserpes[1],
Cecilia Tonelli[2], Valerio Senni[2], and Fabio Federici[2]

[1] Department of Informatics and Telematics, Harokopio University,
Athens 177 78, Greece
ktriakos@hua.gr
[2] Applied Research and Technology, Collins Aerospace, Rome, Italy

Abstract. Compliance data consists of manufacturing quality measures collected in the production process. Quality checks are most of the times computationally expensive to perform mainly due to the amount of collected data. Having trusted solutions for outsourcing analyses to the Cloud is an opportunity for reducing costs of operation. However, the adoption of the Cloud computation paradigm is delayed for the many security risks associated with it. In the use case we consider in this paper, compliance data is very sensitive, because it may contain IP-critical information, or it may be related to safety-critical operations or products. While the technological solutions that protect data in-transit or at rest has reached a satisfying level of maturity, there is a huge demand for securing data in-use. Homomorphic Encryption (HE) is one of the main technological enablers for secure computation outsourcing. In the last decade, HE has reached maturity with remarkable pace. However, using HE is still far from being an automated process and each use case introduces different challenges.

In this paper, we investigate application of HE to the described scenario highlighting, in particular, the main operations of the comparison algorithm, we identify the challenges that HE technology introduces and we propose a solution per challenge. Finally, we evaluate our proposals using one of the open source HE libraries, SEAL, for our implementations.

Keywords: Homomorphic Encryption · Compliance data · Cloud security

This work was supported by the project COLLABS, funded by the European Commission under Grant Agreements No. 871518. This publication reflects the views only of the authors, and the Commission cannot be held responsible for any use which may be made of the information contained therein.

1   Introduction

In recent years, more and more highly regulated manufacturing companies and organizations have outsourced their data and services to a cloud provider[1]. As public or hybrid Cloud-based infrastructures become more and more adopted, there is an increase of threats related to multi-tenancy, further increasing risks of data leakage or compromise. In most of the cases, confidentiality of data is of supreme importance, either due to legislation and compliance obligations or due to the importance of data secrecy for the organization (IP protection needs, competitiveness, international regulations on technical data).

Compliance Data consists of manufacturing quality measures (such as tests results, geometrical measures, images, ...) collected in the production process either by sensors deployed on the industrial equipment itself or by dedicated machines used in intermediate steps of the production (e.g., tools for visual parts inspection). Such data is an essential company asset[2]: it is used as evidence of parts' manufacturing quality as well as to assess responsibilities in case of failures. Since quality checks may be computationally expensive to perform, due to the amount of collected data, having trusted solutions for sharing and processing compliance data is an opportunity for (1) continuously monitoring and improving quality of products through the help of external services, and (2) reducing costs of ownership and operation of an adequate computation infrastructures by outsourcing analyses to the Cloud. However, this collaboration paradigm is also experiencing a delayed adoption for the many security risks associated with it. Indeed, compliance data is very sensitive because it may contain IP-sensitive or competitive information, sometimes related to safety-critical systems.

Therefore, sensitive data should only be shared under the assumption that reliable security guarantees are in place, in particular for what concerns confidentiality. While the technological solutions that protect data in-transit or at rest has reached a satisfying level of maturity, there is a huge demand for securing data in-use. Three are the main technological enablers for secure computation outsourcing: (1) secure multi-party computation (MPC) cryptographic schemes, (2) hardware-based trusted execution environments (TEE), like the Intel SGX or the ARM TrustZone, and (3) Homomorphic Encryption (HE). MPC is a cryptographic tool that allows multiple parties to evaluate a function over their inputs without revealing their individual data. A TEE enables the creation of a protected environment for sensitive data storage and protected execution of critical code. Data access policies are enforced at the CPU level, thus providing data confidentiality and integrity as well as code confidentiality and integrity without relying on the operating system, which may be subject to possible vulnerability or misconfiguration. The concept of HE was introduced in 1978 [17] and it is a

---

[1] Motivations can be found at https://internationaldataspaces.org/wp-content/uploads/dlm_uploads/IDSA-Position-Paper-Data-Sovereignty%E2%80%93Critical-Success-Factor-for-the-Manufacturing-Industry.pdf.

[2] https://www.plex.com/blog/10-strategies-making-compliance-competitive-advantage-aerospace-and-defense.

form of cryptography that enables users to perform computations on encrypted data without first decrypting it, thus providing full confidentiality guarantees on the shared data but no code confidentiality and no guarantees of integrity of data and code. HE research has been considered a holy-grail quest for cryptographers until 2009 and the seminal work of Gentry [12], which revived interest in the community and pushed efforts to increase HE maturity. TEEs provide a broader set of security guarantees with respect to HE (e.g., data and code integrity), but they are subject to vulnerabilities that may reduce the confidentiality guarantees (CVE contains several vulnerabilities affecting Intel SGX[3] and ARM TrustZone[4]) and offer a very limited environment for execution of complex algorithms (e.g., offering a limited amount of memory or lacking numerical libraries to support complex applications like the one presented in this paper). On the other side, HE offers confidentiality guarantees based on the mathematical properties of the underlying crypto primitives implemented in libraries developed by professional cryptographers and widely used in the community, but it may not support all of the algorithm mathematical operations natively, requiring either to reformulate them in terms of the supported mathematical operations or rely on a trusted third party only for those specific operations. In principle, therefore, these technologies offer complementary capabilities and can be fruitfully used in cooperation.

In this paper, driven by the stronger confidentiality guarantees offered by HE, we explore the applicability of HE to a scenario that it is very common in safety-critical parts manufacturing compliance. The reason we choose HE rather than MPC is because there is no shared data among multiple parties, since there is only one client in our use case. Our reference scenario requires the comparison of 3D images to identify any manufacturing differences between the 3D engineering model required for production and the 3D scan of the actual part. We first describe the logic of the 3D comparison algorithm. Then, we identify the challenges that HE technology introduces and we propose solutions for several of those challenges. Finally, we evaluate our proposed solution using one of the open source HE libraries, SEAL[5]. In terms of next steps, we describe opportunities of synergy with other technologies that offer confidential computing capabilities and may help fill the gaps identified in our application of HE.

2   Background

2.1   Homomorphic Encryption Schemes

HE schemes enable data owners to outsource their computations without ever decrypting their data. The data owner is producing a set of keys, a secret key for the decryption of the computations output, a public key for data encryption and several evaluation keys. The evaluation keys are sent to the computation

---

[3] https://cve.mitre.org/cgi-bin/cvekey.cgi?keyword=SGX.

[4] https://cve.mitre.org/cgi-bin/cvekey.cgi?keyword=trustzone.

[5] https://www.microsoft.com/en-us/research/project/microsoft-seal/.

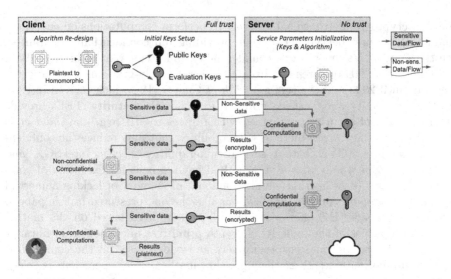

Fig. 1. The HE based outsourcing scenario.

service provider and they are used together with the homomorphic ciphertexts to perform the computations. In the optimal solution, all (HE) operations are computed by the server provider. In the realistic scenario, since some of the HE computations may cause the complexity to grow significantly for the server, the service provider usually performs only part of them, while the rest of the (plaintext) operations are performed by the secret key owner. The HE based computations outsourcing scenario appears in Fig. 1.

The HE schemes support both Boolean and arithmetic operations. Schemes, like FHEW [9] and TFHE [8], are able to compute binary operations and they are perfect choices for non-polynomial functions, such as number comparison. On the other hand, arithmetic schemes, like BGV [5], BFV [4,11] and CKKS [7], can efficiently compute polynomial functions. Of particular interest is the CKKS scheme that supports approximate arithmetic operations, while it takes real or complex numbers as input. CKKS has been proven a very good choice for Machine Learning (ML) applications.

All these HE schemes are based on the hardness of the so-called Learning With Errors (LWE) problem and its generalization Ring LWE (RLWE) problem [15]. The plaintext is a vector of integers (LWE based schemes) or a polynomial (RLWE schemes) with modular coefficients. The plaintext is protected by adding a noise vector from a carefully chosen distribution. The ciphertext noise level must be greater than a minimum level B_0 to be secure and lower than a maximum B_{max} to be decryptable. When it is is greater than B_{max}, decryption is erroneous.

The efficiency of RLWE schemes has been improved by using the data batching [19]. Multiple messages can be packed into a single plaintext, while the corresponding ciphertext supports a simulation of Single Instruction Multiple Data (SIMD) type of computations. In other words, each operation performed on the ciphertext is applied simultaneously to all the packed messages.

2.2   Related Work

The neural-network inference has been identified as the main application area for FHE schemes and there is an extensive line of work demonstrating the capabilities of FHE for privacy preserving machine learning. Hence, most of the HE compilers described in [20], such as CHET and EVA, are designed to address the need of easier FHE programming focusing on the machine learning applications. To the best of our knowledge there is no other work related to manufacturing compliance checks.

2.3   HE Challenges and Limitations

Transforming an algorithm to an HE algorithm is not a straightforward procedure. There are some compilers that can automate this transformation, but they either work only for specific use cases, like ML inference, or they are still not optimal [20]. A human expert is still needed and she has to deal with various challenges. More precisely:

Noise Management: The level of noise increases with each operation and the accumulated noise must remain under the maximum level B_{max} in order the result to be decryptable. The HE algorithm must have the minimal possible multiplicative depth (number of consecutive multiplications) to minimize the noise increase rate. However, when the noise level reaches B_{max}, the ciphertext must be decrypted and re-encrypted to reduce the noise level back to the initial minimum level B_0. This procedure is performed either by a trusted party that has the decryption secret key (the data owner or a TEE module controlled by the data owner) or homomorphically by the server using a function called bootstrapping. Bootstrapping adds very large computation overhead, and it is usually avoided.

Amortization of Storage and Computation Complexity: An HE ciphertext can be seen as a vector several hundred of times larger than a single input message, leading to storage blow-up. To deal with this problem, HE schemes based on RLWE are able to use all the vector slots to store (packing) several different messages in one ciphertext. The operations applied to this ciphertext are performed to all the packed messages simultaneously, simulating an SIMD architecture. The packing strategy can significantly affect the performance of the HE algorithm and it is up to the HE designer to chose the optimal per use case.

Non-polynomial Function Computation: RLWE based schemes that support SIMD are efficient for polynomial function computation, but they perform poorly with non-polynomial ones, like number comparison, square root or division. Three solutions have been proposed: (1) non-polynomial functions are approximated by polynomial ones, at the expense of accuracy [13], (2) switching between HE schemes in the same program, using Boolean or arithemtic HE schemes depending on the function [3,14], but switching is very expensive, and (3) non-polynomial functions are computed in plaintext by a trusted party. In the last solution, either the data owner, with some communication overhead, or a trusted hardware at the server controlled by the data owner, like a TEE module, decrypts the HE ciphertext, performs the function and then HE re-encrypts it.

Branching: Implementing "if then else" conditions or "switch/case" type of operations is very expensive for HE algorithms. In order to hide which branch is computed, the algorithm must compute all possible cases. For instance, see Fig. 2 for a three conditions case. The HE program must homomorphically compute all three functions f_i, $i = 1, 2, 3$.

Algorithm 2.1: Plaintext	Algorithm 2.2: HE computation
if $(cond_1 == 1)$ **then**	$c_1 = HE.f_1$
$\quad c = f_1$	$c_2 = HE.f_2$
else if $(cond_2 == 1)$ **then**	$c_3 = HE.f_3$
$\quad c = f_2$	$c = HE.mult(Enc(cond_1), c_1) +$
else if $(cond_3 == 1)$ **then**	$HE.mult(Enc(cond_2), c_2) \quad +$
$\quad c = f_3$	$HE.mult(Enc(cond_3), c_3)$

Fig. 2. "If then else" conditions (3 conditions example).

3   Point Cloud Mesh Distance in Manufacturing

3.1   COLLABS Use Case

Among all the use cases in manufacturing industry that might benefit from execution of computations directly over encrypted data (see [2,16,21], and [6]), this section focuses on a particular one, that may apply to multiple manufacturing domains and has a well-defined mathematical formulation. The scenario is about manufactured parts compliance, requiring comparison of different 3D images to identify any manufacturing errors. This scenario has been formulated in the context of the COLLABS EU project, working on security in collaborative manufacturing. The scenario has been driven by the quality assurance needs in aerospace parts manufacturing. The purpose of this work is to point out improvements that could be applied, possibly leveraging cutting edge technologies, to enable the secure sharing of confidential data. The COLLABS use case to run compliance checks on manufactured objects is described hereafter in more details.

The actual implementation of the scenario, that is depicted in the left part of Fig. 3, shows the flow of artifacts, highlighting in particular which ones are sensitive and which are not. As mentioned, the limits of this scenario are at least (1) in-house infrastructure is costly to set up and maintain (trust in cloud-based infrastructures), and there is (2) unnecessary circulation of sensitive data. Engineers define a 3D production model, typically a CAD-like file, representing the

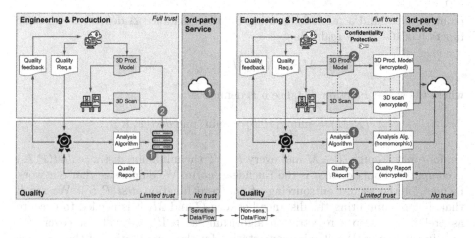

Fig. 3. Compliance checks implementation scenario.

part that shall be manufactured. The 3D model is used to program manufacturing machines, such as CNC Milling machines. Each part, created from these models, is scanned for quality checks, producing a point cloud file[6]. To execute quality checks, every point cloud file and corresponding 3D model file, representing the actual manufactured object and the desired manufactured object respectively, are compared using an industrial metrology tool. Metrology tools implement several algorithms for computing the distance between a scanned object and its reference model, i.e. the distance between a point cloud and a mesh, and provide as output a visualization of the regions where the two models differ, depending on the normal distance from the actual point with respect to expected point, colored following a graduated scale. At the moment, since data is confidential and potentially subject to national regulations for its disclosure, the possibility to use an external cloud infrastructure would pose a cyber-security problem, therefore computations are performed in-house. This paper focuses on the possibility to set up a "HE-enabled cloud scenario"; see right part of Fig. 3. In this case, it is needed to (1) Re-design the analysis algorithm to support HE, (2) HE-Encrypt the 3D production model and the 3D scan accordingly, and (3) HE-Decrypt the Quality report.

3.2 Point Cloud to Mesh Distance Computation and Implementation

To address the automatic compliance check scenario in a formal and mathematical way, we need to introduce some definitions.

Let a point P be an element of $\mathbb{R}^3$, let a point cloud C be a finite set of points of $\mathbb{R}^3$, let a triangle T_i be 3-ple of points in $\mathbb{R}^3$ and let a mesh M be a finite set of triangles. C and M can therefore be represented as a $|C| \times 3$ and a $|M| \times 9$ matrices of real numbers, respectively, where each entry is a point coordinate in the 3 dimensional space. The algorithm to compute the distance between a

---

[6] https://en.wikipedia.org/wiki/Point_cloud.

point cloud C and a mesh M is using function distance Δ defined as a vector function of $|C| = N_C$ real values

$$\Delta(C, M) = [\delta(P_1, M), \ldots, \delta(P_{N_C}, M)]$$

where the scalar function distance δ is defined as

$$\delta(P, M) = \min_{T \in M} d(P, T) = \min_{T \in M} \left(\min_{K \in T} dist(P, K) \right),$$

i.e. for every triangle $T \in M$ and every $K \in T$ the minimum distance $dist(P, K)$ is computed. Function $dist$ is the Euclidean norm. We emphasize that our work focuses on the secure outsourcing of the core function $dist(P, K)$. We recall that, before computing the distance between C and M, it is needed to execute as preliminary step a registration algorithm (as ICP); we will not cover this part in this paper. Possible implementation for the computation of the distance between a point and a mesh is described in [10]. Nevertheless, we implemented a slightly different but equivalent algorithm, which will be referred as the *geometric plaintext* algorithm implementation. Starting from this version of the algorithm, the HE version has been developed.

In order to avoid computing $d(P, K)$ for all $K \in T$, for a triangle $T \in M$, the plane $\pi(T)$ containing T is partitioned in 7 regions. The abstract steps of the algorithm appear in the second column of Table 1. The key point to stress is that the *plaintext* algorithm relies on several Euclidean distance computations (therefore, square root computations), some divisions, comparisons and a branching to state in which of the seven regions triangle T partitions plane $\pi(T)$ every point P stands. The type of operations per step appears in Table 1. We have hidden the details of each step to emphasize the HE challenges.

Table 1. The main operations of the distance algorithm per step.

Steps	Algorithm step for function $d(P, T)$	Type of plaintext operation	HE algorithm
Step 1	Compute vector $\hat{n} = \frac{n}{||n||}$ normal to plane $\pi(T)$, containing T	Real number additions Real number multiplications Real number division Real number square root	All operations HE performed by the server Scale to avoid division and square root Compute scale value
Step 2	Compute the projection of P to plane $\pi(T)$, using $\hat{n}$	Real number additions Real number multiplications	All operations HE performed by the server
Step 3	Compute the projection region Rg of plane $\pi(T)$ using $\hat{n}$	Real number additions Real number multiplications Real number comparison	Operations performed by the Client
Step 4	Depending on region Rg perform Euclidean distance computations to compute d	Real number additions Real number multiplications Real number division Real number square root If branching	All operations HE performed by the server Scale to avoid division and square root Compute scale value Branching (with and without leakage)
Step 5	Return d	Return d	Return d' and scale value The client computes $d = d'/scale$

4 HE Based Module Design

In the setup phase, the data owner creates the HE parameters, the keys and encrypts the data. Data consists of triangles T and points P. Each triangle is defined by 3 points $v_1 = (x_1, y_1, z_1)$, $v_2 = (x_2, y_2, z_2)$ and $v_3 = (x_3, y_3, z_3)$ and the point P is given by $v_p = (x_p, y_p, z_p)$, i.e. in total 9 real numbers per point/triangle pair. We have made the following HE design decisions.

Security Model: We assume that the server owner is honest but curious, i.e. she executes the operations correctly, but she wants to reveal any information that she can. Thus, we focus on confidentiality and not integrity.

Scheme Selection: We have used the CKKS scheme, as the inputs are real numbers and the computation can tolerate approximations. CKKS is a very efficient HE scheme that supports the SIMD computations and allows packing several values in the same plaintext.

SIMD Data Packing: In the encoding phase the (vector) message data is transformed into a polynomial of degree n and it is possible to store up to $\frac{n}{2}$ values in the same plaintext, we refer to them as slots. We propose two different data packing techniques depending on the available workload, the **coordinate-wise packing** and **point-wise packing**. The two packing strategies are analyzed in Sect. 5.

Non-polynomial Functions Computation: Ideally, all computations are meant to be performed at the server-side, only returning the results to the client. However, the distance algorithm uses several non-polynomial functions (real number divisions, comparisons, square roots) that are expensive to be homomorphically computed using CKKS scheme. Thus, we decided to implement those operations using a trusted party. There are two possible options for this: (1) leveraging the data owner (client) or (2) leveraging a TEE-based trusted computation environment controlled by the data owner at the server-side. For the sake of validating proposed algorithm we consider the following approach: a fraction of the original data is sent as cipher-text to the data owner, she performs the required computation in plain-text after decrypting, and she returns a new cipher-text to the server to continue the HE algorithm computation. As a consequence, the HE algorithm requires some extra communication round-trips between the server and data owner. Clearly, this is not an optimal solution, nor the final one: for this reason it does not make sense to evaluate the temporary overhead of the client because it will eventually be eliminated. For the moment, in order to minimize the necessary round-trips and related overhead, we have modified the initial plain-text algorithm to an equivalent version. This is possible by using a scaled version of the computations, i.e. the division and square root operations are postponed for a latter step. At the end, the data owner (client) performs a single square root operation and a division by the homomorphically computed scale number. For future work, in order to avoid all round-trips and eliminate all overhead, we will investigate the use of the TEE at the server-side, as discussed in Sect. 6. In this manner, all the computation will be performed server-side, as expected.

Practical Branching: The HE computation of branching can be very expensive. In our case, depending on the point/triangle input pair, the point P is projected to one of 7 regions of surface $\pi(T)$. Depending on the region, a different Euclidean distance is computed. In a use-case-agnostic implementation, all 7 functions must be computed homomorphically in order to hide the projection result from the server (see Fig. 2), resulting in an inefficient implementation. However, for the specific use-case, the projection region of each triangle is not subject to confidentiality and, if leaked, doesn't provide any significant information to a malicious actor, while the cost of hiding it is too high. This peculiarity gives us another alternative to the TEE for addressing this gap: leaking the information for better performance, e.g. either by performing the computation server-side or client-side. Information leakage can be formally defined, as it was demonstrated for AI oblivious inference [18]. Cryptography has always been a trade-off between security and practicality and in several cases the protection cost of information's secrecy is disproportionately high compared to the protected information's importance. In our case, we decided to leak this information to gain efficiency. The proposed solution appears in Fig. 4. The region R_g is leaked, but only one function is computed. Regarding the efficiency, the runtime is improvement is proportional, i.e. the computational overhead of the branching is reduced by a factor of 7 (averaging over 1000 trials we have a reduction from 0.9 s to 0.1 s).

Algorithm 4.1: Plaintext

Algorithm 4.2: HE computation with leakage

if (R_g == 1) then
   $c = f_1$
else if (R_g == 2) then
   $c = f_2$
else if (R_g == 3) then
   $c = f_3$
else if (R_g == 4) then
   $c = f_4$
else if (R_g == 5) then
   $c = f_5$
else if (R_g == 6) then
   $c = f_6$
else if (R_g == 7) then
   $c = f_7$

if (R_g == 1) then
   $c = HE.f_1$
else if (R_g == 2) then
   $c = HE.f_2$
else if (R_g == 3) then
   $c = HE.f_3$
else if (R_g == 4) then
   $c = HE.f_4$
else if (R_g == 5) then
   $c = HE.f_5$
else if (R_g == 6) then
   $c = HE.f_6$
else if (R_g == 7) then
   $c = HE.f_7$

Fig. 4. Depending on the region R_g a different Euclidean distance-like function f_i is computed. When the value of R_g is leaked, only one of them is homomorphically computed.

5   Packing Strategies Evaluation

One of the main features of RLWE-based HE schemes that extremely improves performance is data packing, i.e. the ability to store multiple values in a single plaintext. We propose two data packing techniques for our scenario, the coordinate-wise and point-wise packing. In this section, we analyze the two packing strategies and evaluate the storage overhead.

The coordinate-wise packing (CWP) is a straightforward technique used by all the HE compilers. Each coordinate, x_i, y_i and z_i, $i = \{1, 2, 3, p\}$ is encoded in a separate plaintext and then encrypted, i.e. 12 ciphertexts are used for each point/triangle pair. These 12 ciphertexts can store, and process in parallel (SIMD), up to $\frac{n}{2}$ pairs, if these pairs are available as a batch input. This packing approach appears in Fig. 5.

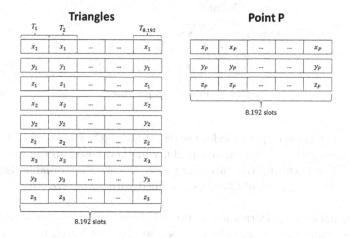

Fig. 5. Coordinate-wise packing.

Although packing multiple data in a single plaintext increases the efficiency, there are several limitations. We are not able to perform operations to a subset of the plaintext's slots and we can have random access to a single slots (like in a array data structure). However, cyclic rotations of the slot values are allowed in the plaintext. In the point-wise packing (PWP) technique, we take advantage of the rotations to reduce the amount of ciphertexts. In more detail, each point v_i, $i = \{1, 2, 3, p\}$ is encoded in a different plaintext and 4 ciphertexts are produced. Since rotations are included in our computations, each point's coordinates are repeated 3 times and in total, 9 slots are required for each point/triangle pair as shown in Fig. 6. Thus, $\frac{1}{9} \cdot \frac{n}{2} = \frac{n}{18}$ point/triangle pairs can be processed in parallel. Figure 7 illustrates the packing method. It is clear that per point/triangle pair 6 slots are filled with repeated coordinate values. We refer to them as *redundancy slots*.

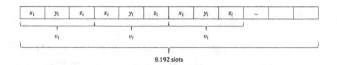

Fig. 6. Plaintext in Point-wise packing.

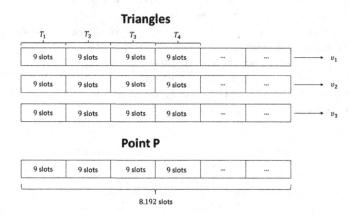

Fig. 7. Point-wise packing.

The data owner is responsible for the data batching before the encoding phase and for choosing the most adequate packing method based on the workload, as shown in Table 2. During the encoding phase, in case the available data is less than the $\frac{n}{2}$ slots, the rest of the slots are filled with zeros. We refer to them as *idle slots*.

We evaluate the performance of the HE-based program, for security level of 128-bits. The CKKS parameters are selected by the SEAL library following the HE standard [1]. The polynomial degree is chosen $n = 16.384$ and number of messages that can be stored in a ciphertext (number of available slots) is $\frac{n}{2} = 8.192$.

Table 2. The number of ciphertexts used per packing strategy as a function of the available input point/triangle pairs.

Input Pairs	$9,1 \cdot 10^2$	$1,82 \cdot 10^3$	$2,73 \cdot 10^3$	$3,64 \cdot 10^3$	$4,55 \cdot 10^3$	$5,46 \cdot 10^3$	$8,192 \cdot 10^3$	10^4
PWP	4	8	12	16	20	24	40	44
CWP	12	12	12	12	12	12	12	24

Regarding the storage overhead, in the PWP strategy, multiple of 4 ciphertexts are used and every 4 ciphetexts can store up to 910 point/triangle pairs, while using CWP, groups of 12 ciphertexts are used and each group can store up to 8.192 point/triangle pairs. In case 2.730 point/triangle are available, PWP is

Table 3. The number of ciphertexts used in hybrid strategy as a function of the available input point/triangle pairs.

Input Pairs	910	8.192	9.102	10.012	10.992
Hybrid	–	–	16	20	24

Table 4. The number of redundancy and idle slots per plaintext of packing strategies as a function of the available input point/triangle pairs.

Input Pairs		$9,1 \cdot 10^2$	$1,82 \cdot 10^3$	$2,73 \cdot 10^3$	$8,192 \cdot 10^3$
Redundancy	PWP	5.460	5.460	5.460	5.460
	CWP	0	0	0	0
Idle	PWP	2	2	2	2
	CWP	7.282	6.372	5.462	0

the best strategy since for the same amount of ciphertexts there are less idle slots. In order to optimize the ciphertext usage, we must have workloads, i.e. available pairs for comparison, that are multiple of 910 point/triangle pairs. The number of the ciphertexts as a function of the workload appears in Table 2. It is clear that PWP requires significantly fewer ciphertexts than CWP. Although, PWP technique is suitable for low data workloads, the total number of redundancy and idle slots shown in Table 4 should also been taken into account.

In case that the available point/triangle pairs are more than 8.192, but less than 10.922, a hybrid strategy can be used by combining both packing techniques. The idea is using the CWP for the first 8.192 pairs and PWP for the other 2.730, in order to have the minimum number of idle slots and ciphertexts. If the available pairs are less than 8.192 there is no need to use the hybrid version. The number of required ciphertexts using hybrid strategy as a function of the workload appears in Table 3.

6   Conclusion and Future Work

In this paper, we investigated the use of HE based computation in manufacturing. More precisely, we demonstrated how to securely outsource comparison of 3D images and, specifically, the comparison of the 3D engineering model required for production with the 3D scan of the actual part. We identified the challenges that HE technology imposes, we introduced solutions for some of those challenges. We have redesigned the plaintext algorithm to be HE friendly, by postponing the computations of non-polynomial functions to a latter step. We also propose leaking insignificant information to the server and take advantage of the SIMD, by introducing two packing methods, in order to increase the efficiency. For our proof-of concept implementations we used the open source HE libraries, SEAL.

As anticipated in the introductory part of this paper, HE is not able to address all the operations required by the compliance algorithm, thus forcing the solution to rely on additional network transfers of portions of data to rely on the trusted client environment for performing the unsupported operations. Despite not being able to quantify at this stage the cost of these extra network communications and the client cost for performing the unsupported operations, this is certainly an unacceptable solution for scalability and industrial use. For this reason, the immediate next step consists in extending the solution architecture to incorporate server-side TEE capabilities to host the plain-text operations that are currently offloaded back to the client. This would be a more realistic setup for three reasons: (1) the abolition of unnecessary resource consumption on the network and client sides, (2) the extension to the currently unsupported operations in the compliance algorithm, and (3) the use of the TEE only on a subset of data (thus being within more limited memory requirements) and on a subset of mathematical operations (thus potentially requiring a minimal set of additional libraries to be added to the TEE secure domain). We also envision to explore alternative HE crypto schemes and comparison of our manually-tuned algorithm with an implementation produced by HE compilers like Microsoft's EVA[7].

References

1. Homomorphic encryption standardization. https://homomorphicencryption.org/standard/ (2018)
2. Abdallah, H.B., Orteu, J., Jovancevic, I., Dolives, B.: Three-dimensional point cloud analysis for automatic inspection of complex aeronautical mechanical assemblies. J. Electron. Imaging **29**(4), 041012 (2020). https://doi.org/10.1117/1.JEI.29.4.041012
3. Boura, C., Gama, N., Georgieva, M., Jetchev, D.: CHIMERA: Combining ring-LWE-based fully homomorphic encryption schemes. J. Math. Cryptology **14**(1), 316–338 (2020)
4. Brakerski, Z.: Fully homomorphic encryption without modulus switching from classical GapSVP. In: Safavi-Naini, R., Canetti, R. (eds.) CRYPTO 2012. LNCS, vol. 7417, pp. 868–886. Springer, Heidelberg (2012). https://doi.org/10.1007/978-3-642-32009-5_50
5. Brakerski, Z., Gentry, C., Vaikuntanathan, V.: (Leveled) fully homomorphic encryption without bootstrapping. ACM Trans. Comput. Theory (TOCT) **6**(3), 1–36 (2014)
6. Chen, X.: An innovative detection method of high-speed railway track slab supporting block plane based on point cloud data from 3D scanning technology. Appl. Sci. **9**(16), 3345 (2019)
7. Cheon, J.H., Kim, A., Kim, M., Song, Y.: Homomorphic encryption for arithmetic of approximate numbers. In: Takagi, T., Peyrin, T. (eds.) ASIACRYPT 2017. LNCS, vol. 10624, pp. 409–437. Springer, Cham (2017). https://doi.org/10.1007/978-3-319-70694-8_15
8. Chillotti, I., Gama, N., Georgieva, M., Izabachène, M.: Faster fully homomorphic encryption: bootstrapping in less than 0.1 seconds. In: Cheon, J.H., Takagi, T. (eds.) ASIACRYPT 2016. LNCS, vol. 10031, pp. 3–33. Springer, Heidelberg (2016). https://doi.org/10.1007/978-3-662-53887-6_1

---

[7] https://github.com/microsoft/EVA.

9. Ducas, L., Micciancio, D.: FHEW: bootstrapping homomorphic encryption in less than a second. In: Oswald, E., Fischlin, M. (eds.) EUROCRYPT 2015. LNCS, vol. 9056, pp. 617–640. Springer, Heidelberg (2015). https://doi.org/10.1007/978-3-662-46800-5_24

10. Eberly, D.: Distance between point and triangle in 3D. https://www.geometrictools.com/Documentation/DistancePoint3Triangle3.pdf (1999)

11. Fan, J., Vercauteren, F.: Somewhat practical fully homomorphic encryption. IACR Cryptol. ePrint Arch. **2012**, 144 (2012)

12. Gentry, C.: Fully homomorphic encryption using ideal lattices. In: Proceedings of the Forty-first Annual ACM Symposium on Theory of Computing, pp. 169–178 (2009)

13. Juvekar, C., Vaikuntanathan, V., Chandrakasan, A.P.: GAZELLE: a low latency framework for secure neural network inference. In: Enck, W., Felt, A.P. (eds.) 27th USENIX Security Symposium, USENIX Security 2018, Baltimore, MD, USA, 15–17 Aug 2018, pp. 1651–1669. USENIX Association (2018). https://www.usenix.org/conference/usenixsecurity18/presentation/juvekar

14. Lu, W., Huang, Z., Hong, C., Ma, Y., Qu, H.: PEGASUS: bridging polynomial and non-polynomial evaluations in homomorphic encryption. In: 42nd IEEE Symposium on Security and Privacy, SP 2021, San Francisco, CA, USA, 24–27 May 2021, pp. 1057–1073. IEEE (2021). https://doi.org/10.1109/SP40001.2021.00043

15. Lyubashevsky, V., Peikert, C., Regev, O.: On ideal lattices and learning with errors over rings. In: Gilbert, H. (ed.) EUROCRYPT 2010. LNCS, vol. 6110, pp. 1–23. Springer, Heidelberg (2010). https://doi.org/10.1007/978-3-642-13190-5_1

16. Nguyen, C.H.P., Choi, Y.: Comparison of point cloud data and 3D CAD data for on-site dimensional inspection of industrial plant piping systems. Autom. Constr. **91**, 44–52 (2018). https://doi.org/10.1016/j.autcon.2018.03.008. https://www.sciencedirect.com/science/article/pii/S0926580517308221

17. Rivest, R.L., et al.: On data banks and privacy homomorphisms. Found. Secure Comput. **4**(11), 169–180 (1978)

18. Rizomiliotis, P., Diou, C., Triakosia, A., Kyrannas, I., Tserpes, K.: Partially oblivious neural network inference. In: di Vimercati, S.D.C., Samarati, P. (eds.) Proceedings of the 19th International Conference on Security and Cryptography, SECRYPT 2022, Lisbon, Portugal, 11–13 July 2022, pp. 158–169. SCITEPRESS (2022). https://doi.org/10.5220/0011272500003283

19. Smart, N.P., Vercauteren, F.: Fully homomorphic SIMD operations. IACR Cryptology ePrint Archive, Paper 2011/133 (2011). http://eprint.iacr.org/2011/133

20. Viand, A., Jattke, P., Hithnawi, A.: Sok: fully homomorphic encryption compilers. In: 42nd IEEE Symposium on Security and Privacy, SP 2021, San Francisco, CA, USA, 24–27 May 2021, pp. 1092–1108. IEEE (2021). https://doi.org/10.1109/SP40001.2021.00068

21. Xu, Z., Kang, R., Lu, R.: 3D reconstruction and measurement of surface defects in prefabricated elements using point clouds. J. Comput. Civ. Eng. **34**(5), 04020033 (2020). https://doi.org/10.1061/(ASCE)CP.1943-5487.0000920

Author Index

Printed in the United States
by Baker & Taylor Publisher Services